WASTEWATER BIOSOLIDS TO COMPOST

HOW TO ORDER THIS BOOK

BY PHONE: 800-233-9936 or 717-291-5609, 8AM–5PM Eastern Time

BY FAX: 717-295-4538

BY MAIL: Order Department
Technomic Publishing Company, Inc.
851 New Holland Avenue, Box 3535
Lancaster, PA 17604, U.S.A.

BY CREDIT CARD: American Express, VISA, MasterCard

BY WWW SITE: http://www.techpub.com

WASTEWATER BIOSOLIDS TO COMPOST

Frank R. Spellman, Ph.D.

Environmental Health & Safety Manager
Hampton Roads Sanitation District
Virginia Beach, Virginia

LANCASTER · BASEL

Wastewater Biosolids to Compost
a **TECHNOMIC**® publication

Published in the Western Hemisphere by
Technomic Publishing Company, Inc.
851 New Holland Avenue, Box 3535
Lancaster, Pennsylvania 17604 U.S.A.

Distributed in the Rest of the World by
Technomic Publishing AG
Missionsstrasse 44
CH-4055 Basel, Switzerland

Printed in the United States of America
10 9 8 7 6 5 4 3 2 1

Main entry under title:
Wastewater Biosolids to Compost

A Technomic Publishing Company book
Bibliography: p.
Includes index p. 243

Library of Congress Catalog Card No. 96-60894
ISBN No. 1-56676-461-0

To Danny Finley
Chief Operator, Peninsula Composting Facility
Newport News, Virginia

Table of Contents

Preface

SIMPLY stated: This text takes the view that there is a more reasonable approach to choose in the ultimate management of wastewater biosolids: *composting*. Moreover, this view holds that biosolids should be regarded as a valuable resource that can be reused or recycled.

The beginning of the modern environmental movement can be dated from Earth Day, in April of 1970. Since about the same time, the conversion of wastewater biosolids to compost has grown as a beneficial reuse process.

Composting biosolids has grown in popularity for other reasons. To begin with, when ocean dumping of biosolids was banned in the United States and other countries, another means of disposal for biosolids had to be found. Incineration was one option; however, with increasingly more stringent regulations regarding air quality, incineration of biolsolids is no longer the prime option that it was in the past. Biosolids have also been deposited in landfills or as a cover for the fill. The problem with that approach is that it is difficult to find landfill space for present and future disposal. While it is true that land reclamation still holds promise for future management of wastewater biosolids, it is also true that most of the land to be reclaimed is isolated from the larger Wastewater Treatment Plants (WWTPs) that serve metropolitan areas.

For all of these reasons, composting biosolids has gained significantly as a management choice.

This text is designed to provide the decision-maker, site manager, site engineer, as well as local public administrators with a user-friendly guidebook that not only explains the process but also explains how to build a composting facility—all presented in fundamental/basic language.

In addition to providing the reader with a guide for planning, this text also covers EPA 503 regulations, testing procedures, advancements in odor control, marketing of the product, and the economics of composting.

This text not only sends the message that composting biosolids is a beneficial reuse process, but also points out that composting biosolids is not the "pie-in-the-sky" solution to all biosolids management problems. Particular attention is given to cost. It is also pointed out that composting biosolids is not the only option available for beneficial reuse of biosolids. In some cases, it may be more cost effective to co-compost with Municipal Solid Waste (MSW) or paper waste. This option especially increases in importance whenever siting problems for a new or existing composting facility cannot be overcome.

The main point is that composting biosolids is not a money-making enterprise. Instead, it is a viable option that can be used to reuse biosolids in an earth-friendly manner. Isn't this what waste disposal should be all about?

FRANK SPELLMAN
Virginia Beach, VA

Acknowledgements

THIS text is a compilation of general information, documented experience, and engineering expertise provided to me by several experts in the wastewater treatment industry.

In particular, I would like to thank James R. Borberg, PE, General Manager, Hampton Roads Sanitation District (HRSD), Virginia Beach, Virginia, who allowed me to access and use HRSD materials in this text. Moreover, several other HRSD personnel supported my efforts in this work. I want to thank the following HRSD personnel: David Waltrip, PE, Director of Treatment; Mardane McLemore, PE, Asst. Director of Treatment; Rick Baumler, Treatment Manager; Rhonda Oberst, Recycling Manager/Agronomist; Danny Finley, Chief Operator of the Peninsula Composting Facility, Newport News, Virginia; Brian McNamara, PE, Project Engineer; David Morse, Automotive Superintendent; Chuck Lewicki, Plant Manager, Wiliamsburg WWTP; and the Drafting Division for their assistance with graphic arts.

Special thanks are due to Rosalyn Hopkins (my trusted associate), who works hard to keep me out of trouble; no easy task.

CHAPTER 1

Introduction

> It is very difficult to make
> an accurate prediction,
> especially about the future.
>
> —Niels Bohr

A rational person would have difficulty arguing against Bohr's view concerning the difficulty of making accurate predictions about the future. However, rational people are capable of recognizing that, in a few instances, there are exceptions to every rule. More specifically, as the result of certain "actions," knowledgeable and observant individuals can make fairly accurate predictions about the potential future consequences of such actions. As a case in point, consider the actions of Man that pollute the air we breathe, the water we drink, and the land we live on and gain our sustenance from. When contamination and destruction of our life-sustaining enviromment occur on a daily basis, certain fairly accurate predictions can be made about the future consequences. Through the observance and/or the awareness of certain "actions," a sense of foreboding concerning the dire consequences of Man destroying the environment motivated a group of concerned and enlightened individuals to organize the first Earth Day celebration in 1970.

The organizers of the first Earth Day celebration were concerned about certain "actions" of Man. To a degree, the organizers' concern was driven by obvious "predictors" of a quality of life anticipated for the future that was not very promising. Some of these organizers and other concerned individuals made dire predictions based on what they had observed, on what they had read, or on what they had heard. For example, they might have observed or learned about certain rivers within the United States that were so oil soaked that they actually burned. Moreover, others had observed, heard or read about skies above metropolitan areas that were red with soot. Others had breathed air that they could actually see. And still others had observed lakes choked with algae, lakes that were dying.

Then there were the mountains they had observed, had read about, or had heard about. These were mountains unlike the Alps or the Rocky Mountains, however. Instead, these were mountains of trash, garbage, refuse, discarded materials, and other waste products. As with all things that disgust Man, the same fate awaited these mountains of waste and filth; that is, they became unbearable for Man to live with and were torn down. After teardown, the entire unsightly, stinking mess was deposited into rivers, lakes, streams, oceans, or landfills. Sometimes these mountains of unwanted wastes were torn down and piled up again on barges that were towed from port to port to port with no place to land; no one wanted these floating mountains of waste: Not in My Backyard (NIMBY).

All these observations, of course, were indicators of what was occurring environmentally in the here and now; moreover, they were "actions" portending what was in store for the inhabitants of earth and for the future generations to come. Thus, these indicators of environmental problems became predictors of greater environmental problems in the future.

Along with the organizers and participants of that first Earth Day celebration in 1970 there were other citizens who were concerned about their futures and the futures of their loved ones. And although Niels Bohr was correct in his statement about the difficulty of making accurate predictions, especially about those in the future, in 1970 it was clear to many concerned individuals that if corrective actions were not quickly taken to protect and preserve the earth's environment, there would be no need to worry about making future predictions; that is, there would be no future to predict.

To say that we face huge environmental challenges today, as was the case in 1970, is an accurate statement. While it is true that since that first Earth Day, progress has been made in restoring the earth's environment, it is also true that we still have a long way to go before the "predictors" or "indicators" or the future consequences of the on-going damage to earth's environment are less salient than they are today. It can be said with a great deal of accuracy, therefore, that the quality of life here on earth is directly connected to our "actions."

However, not all the news concerning man-made waste and its disposal is of the doom-and-gloom variety. For example, it is noteworthy to take into account the steps that have been taken in recent years to clean up our air and our lakes, and to properly dispose of our waste using earth-friendly disposal techniques. One such clean-up step is described in the following case study, which was presented at WEFTEC '95 (Miami) by Oberst and Robinson (1995) and details an example of beneficial reuse of biosolids; namely, Hampton Roads Sanitation District's biosolids ash recycling program.

CASE STUDY

Hampton Roads Sanitation District (HRSD) is located in southeastern Virginia. HRSD operates nine wastewater treatment plants with a capacity of 210.5 million gallons per day, and serving an area of 750 square miles and 1.2 million customers. At six plants, the biosolids is incinerated generating 31,000 cubic yards of biosolid ash annually. Historically, the biosolids has been disposed of in sanitary landfills.

Since 1981, two of the HRSD plants have used their biosolids to manufacture a soil amendment or compost product. As a result, approximately 20,000 cubic yards of compost is produced annually and sold to garden centers and the public. Since 1984, one HRSD plant has land-applied the biosolids generated. Approximately 1,500 acres of farm land in the Hampton Roads area receive annual applications of the biosolids. The plant produces over 3 million gallons of liquid biosolids (8% total solids) and 7,000 dry tons of cake biosolids (20% total solids) annually.

HRSD had solved its ultimate disposal problems with biosolids that were not incinerated, but was left with the problem of disposing of biosolid ash left after incineration. With 31,000 cubic yards of ash generated annually, and with landfill costs ranging between $36–$53 per ton plus transportation costs, HRSD was looking for a beneficial reuse alternative.

After several years of research, pilot studies, and regulatory review, HRSD and its consultants devised a strategy for beneficial reuse of its biosolids ash. This strategy first involved developing a project that focused on a marine application; that is, preventing shoreline erosion. This seemed a logical step because HRSD serves an area that is on and around the Atlantic Ocean. A shoreline erosion-control product was selected for manufacture using HRSD's biosolid ash. According to the U.S. Army Corps of Engineers (1992), the shoreline erosion-control revetment project (concretelike blocks containing 20% biosolid ash) performed as designed.

Several other biosolid ash reuse projects were implemented. One project of note involved using ash as an ingredient in creating a product line called Biocritters (see Figure 1.1). Biocritters are novelty products in the form of turtles, frogs, rabbits, dogs and other "critters." The public has accepted Biocritters. Moreover, these "critters" have helped to tell the story of environmental protection and promote the beneficial reuse of biosolid ash.

HRSD has realized cost savings and other benefits from its biosolid ash reuse program. For example, the need for worker involvement in handling and exposure activities has been reduced, thereby reducing labor and medical costs. The economics of hauling ash has been affected in that instead of HRSD bearing the total cost of hauling the ash to a landfill, the ash being hauled to the factory for reuse in manufacturing products now recovers some of the expense through sales of finished products. Further,

Figure 1.1 "Biocritters," manufactured using biosolid ash from HRSD and advertisement for HRSD Compost product, used by permission.

because the need to landfill ash has decreased significantly, tipping fees have been reduced. In addition to realizing an overall cost reduction of 25% through its ash recycling program, HRSD has been able to conserve valuable landfill space.

The purpose of this text, like the case study just discussed, is to address one small but important and significant aspect (or process) of making man-made waste disposal more earth-friendly: *biosolids composting.* Since 1970, much progress has been made in sewage treatment technology. As a case in point, consider the lakes and streams that were "unswimmable" and "unfishable" just a few decades ago. Corrective actions in treating domestic and industrial wastes have advanced to the point and have been underway for a long enough period now so that today one can visit most local lakes and streams and clearly see the lake or river bottom near a shallow shoreline. In some cases, at local streams and lakes, the posted signs warning that FISHING AND SWIMMING is prohibited have come down. These are examples of an environmental improvement that can be readily seen. This visible improvement is also a "predictor" of what the future can hold if we respect lakes and streams, and thus the environment.

Recent improvements in the water quality of streams and lakes are only a small part of the progress that has been made. Improvements in wastewater technology have also worked to improve the quality of water we use; that is, the water we drink.

This last statement may seem strange to some readers. Some might ask: How does wastewater treatment improve the quality of potable water when we do not receive our drinking water from wastewater treatment plant effluent? That is, some readers know that effluent from wastewater treatment plants is not normally cross-connected with their municipality's drinking water supply. The point is that many communities draw water from streams and rivers for use in domestic potable water supplies, and these same streams and rivers serve as outfalls, normally upstream, for wastewater treatment plant effluent. Communities are growing. Populations within these burgeoning communities are also growing. Along with growth in community size and in population comes a corresponding growth in the need for more potable water. Thus, the stream or river that provides the water supply and serves as the outfall for wastewater treatment plant effluent is put under increasing demand for its main product: potable water.

From the preceding it is evident that wastewater treatment has been instrumental in Man's recent and ongoing effort to preserve the environment. Moreover, wastewater treatment technology has worked to lessen the negative consequences of polluting our water systems.

Wastewater treatment technology does not clean our waters for free, however. In terms of financial costs, wastewater treatment is not inexpensive. In addition, there is another "cost" or problem that must also be addressed, related to the solid waste material that is "left over" after the wastestream has been treated. This wastewater solid waste product (sludge or biosolids) must either be disposed of or reused, and doing either is not inexpensive.

Volumes of research data have been collected and subsequently printed on the disposition of the liquid portion of treated wastewater effluent. However, the wastewater solids portion (sludge or biosolids) of the wastestream has received less attention. These wastewater solids (commonly grouped under the unflattering term *sludge*) and one promising process that enables their safe disposal or reuse are what this text is designed to address. Even though sewage sludge disposal problems are but a small part of our overall waste disposal problem, continued improper disposal of sewage sludge and other wastes could contribute to an imbalance between the environment and the lifeforms it sustains.

Since the early 1970s, attention to the possibilities to be realized from biosolids composting has increased. Biosolids composting is a cost effective and environmentally sound alternative for the stabilization and ultimate disposal of sewage sludge. To put it simply, composting wastewater biosolids is an earth-friendly, beneficial-use, waste disposal technology.

The recent increased tendency to use biosolids composting as an ultimate solid waste disposal process has been spurred on by increasingly stringent air pollution regulations (making incineration illegal or expensive) and waste disposal requirements coupled with the anticipated shortage of landfill space.

Biosolids composting has come a long way from its inception in the early 1970s. For example, Goldstein et al. (1994) point out that a BioCycle survey conducted in 1983 included a total of ninety biosolids composting projects. By 1994 the total had grown to more than 300. Of those, 201 are in full-scale operation in early 1996.

Before entering into a discussion of the biosolids composting process and of designing a facility capable of converting wastewater biosolids solids to compost and, whether or not such a facility is economically feasible, it is important to discuss a few basic concepts that are prevalent throughout this text:

(1) As stated earlier, wastewater solids have traditionally been grouped under the title of *sewage sludge* or sludge. These terms have wide recognition and are used by the regulators and their regulatory documents pertaining to U.S. EPA's 503 Regulations (Standards for the Use or Disposal of Sewage Sludge). At the same time, the terms *sewage sludge* or *sludge* conjure up connotations that do not fit with the value of this waste product. This view can be better understood when one considers the following discussion by Vesilind (1980):

> Sludge, in my opinion, should be thought of as a waste product with value (no contradiction) and a material which should find its proper use, not just a method of disposal . . . to some people, sludge is still an ugly four letter word. To others (and to me) it is a residual of our society which must be treated and used to the maximum benefit of mankind, recognizing that mankind is but one species on this wonderful planet. (pp. 315–316)

Vesilind's point is well taken. Moreover, the argument about changing an ugly four-letter word into something more fitting has resulted in a recent trend to incorporate the use of the term *biosolids* as the common and preferred term in place of sludge. The term *biosolids* is a fitting replacement because it reflects the fact that a significantly large portion of the composition of wastewater solids is composed of biomass—a biomass that can be put to beneficial use.

(2) The model depicted in Figure 1.1 is the paradigm or prototype that will be used in this text to explain the biosolids-to-compost process. The model is one of several models that are available. However, since this is an introductory presentation designed for decision-makers and more concerned with the "process" than the types of composting processes available, the aerated static pile (ASP) process will serve as our model and focus. The model depicted in Figure 1.1 will be called the ASP model. The ASP compost process closely resembles actual online processes that have successfully produced marketable biosolids-derived compost for several years. Moreover, the ASP model has an extensive, well-documented track record (of the 201 biosolids composting sites in operation, more than ninety use the aerated static pile method) (Goldstein et al., 1994).

TERMINOLOGY

Every process or branch of science has its own language for communication. Biosolids composting is no different. In order to work even at the edge of composting, it is necessary to acquire a fundamental vocabulary of the components that make up the process of biosolids composting. Therefore, before we discuss the scope of this text and then move on to specific topics, it will be beneficial for the reader to review the definitions of the key terms used in this text.

Note: Definitions asterisked (*) are taken from U.S. Environmental Protection Agency—*Federal Register,* Volume 58, Friday, February 19, 1993, Rules and Regulations Part 503 Standards for the Use or Disposal of Sewage Sludge. The EPA does not use the term *biosolids* in its rules and regulations; this term is provided by the author and by practitioners in the field. Nonasterisked definitions are taken from U.S. EPA's *Composting Yard & Municipal Solid Waste* (1995).

DEFINITIONS OF KEY TERMS

Aerated static pile—Composting system using controlled aeration from a series of perforated pipes running underneath each pile and connected to a pump that draws or blows air through the piles.

Aeration (for composting)—Bringing about contact between air and composted solid organic matter by means of turning or ventilating to allow microbial aerobic metabolism (biooxidation).

Aerobic—Composting environment characterized by bacteria active in the presence of oxygen (aerobes); generates more heat and is a faster process than anaerobic composting.

Anaerobic—Composting environment characterized by bacteria active in the absence of oxygen (anaerobes).

Bacteria—Unicellular or multicellular microscopic organisms.

**Bagged biosolids*—Biosolids that is sold or given away in a bag or other container (i.e., either an open or a closed vessel containing 1 metric ton or less of biosolids).

Bioaerosols—Biological aerosols that can pose potential health risks during the composting and handling of organic materials. Bioaerosols are suspensions of particles in the air consisting partially or wholly of microorganisms. The bioaerosols of concern during composting include actinomycetes, bacteria, viruses, molds, and fungi.

Biosolids—A term that is not used by the EPA in its 503 regulations, but that has become commonly used in the wastewater treatment industry as a replacement for the term *sewage sludge.* Biosolids are the solid, slime-solid, or liquid residue generated during the treatment of domestic sewage in a wastewater treatment facility. Biosolids includes, but is not limited to, domestic sewage, scum, solids removed during primary, secondary, or advanced treatment processes. The definition of biosolids also includes a material derived from biosolids (i.e., biosolids to compost).

Biosolids composting—The process involving the aerobic biological degradation or bacterial conversion of dewatered biosolids that works to produce compost that can be used as a soil amendment or conditioner.

Biosolids quality parameters—The EPA determined that three main parameters should be used in gauging the biosolids' quality: (1) the relevant presence or absence of pathogenic organisms, (2) pollutants, and (3) the degree of attractiveness of the biosolids to vectors. There can be a number of possible biosolids qualities. To express or describe the biosolids that meet the highest quality for all three biosolid quality parameters, the term *Exceptional Quality* or EQ has come into common use.

**Bulk biosolids*—Biosolids that is not sold or given away in a bag or other container for application to the land.

Bulking agents—Materials, usually carbonaceous such as sawdust or woodchips, added to a compost system to maintain airflow by preventing settlement and compaction of the compost.

Carbon-to-nitrogen ratio (C:N ratio)—Ratio representing the quantity of carbon (C) in relation to the quantity of nitrogen (N) in a soil or organic material; determines the composting potential of a material and serves to indicate product quality.

Compost—The end product (innocuous humus) remaining after the composting process is completed.

Curing—Late stage of composting, after much of the readily metabolized material has been decomposed, which provides additional stabilization and allows further decomposition of cellulose and lignin (found in woodylike substances).

Endotoxins—A toxin produced within a microorganism and released upon destruction of the cell in which it is produced. Endotoxins can be carried by airborne dust particles at composting facilities.

EPA's 503 regulation—In order to ensure that sewage sludge (biosolids) is used or disposed of in a way that protects both human health and the environment, under the authority of the Clean Water Act as amended, the U.S. Environmental Protection Agency (EPA) promulgated, at 40 CFR Part 503, Phase I of the risk-based regulation that governs the final use or disposal of sewage sludge (biosolids).

Exceptional quality (EQ) sludge (biosolids)—Although this term is not used in 40 CFR Part 503, it has become shorthand for biosolids that meet the pollutant concentrations in Table 3 of Part 503.13(b)(3); one of the six class A pathogen reduction alternatives in 503.32(a); and one of the vector attraction reduction options in 503.33(b)(1)–(8) (O'Dette, 1995).

Feedstock—Decomposable organic material used for the manufacture of compost.

Metric ton—One (1) metric ton, or 1,000 kg, equals about 2,205 lbs, which is larger than the short ton (2,000 lb) usually referred to in the British system of units. The metric ton unit is used throughout this text.

**Pathogen organisms*—Specifically, *Salmonella* and *E-coli* bacteria, enteric viruses, or visible helminth ova.

**Pollutant*—An organic substance, an inorganic substance, a combination of organic and inorganic substances, or a pathogenic organism that, after discharge and upon exposure, ingestion, inhalation, or assimilation into an organism either directly from the environment or indirectly by ingestion through the food chain, could, on the basis of information available to the EPA, cause death, disease, behavioral abnormalities, cancer, genetic mutations, physiological malfunctions, or physical deformations in either organisms or offspring of the organisms.

Stability—State or condition in which the composted material can be stored without giving rise to nuisances or can be applied to the soil without causing problems; the desired degree of stability for finished compost is one in which the readily decomposed compounds are broken down and where only the decomposition of the more resistant biologically decomposable compounds remains to be accomplished.

Vectors—Refers to the degree of attractiveness of biosolids to flies, rats, and mosquitoes, that could come into contact with pathogenic organisms and spread disease.

SCOPE OF THE TEXT

This text does not include problem sets and assignments; thus, it is not a textbook. What this text is about is *information.* This text provides information for consumption by decision-makers who might be wastewater operators, municipal managers and/or administrators, and those individuals in the public sector who may be interested in the subject matter. The text has as its main objective to describe, explain, and evaluate a successful biosolids-to-compost option. It is hoped that the text will be interesting and useful not only to those in charge of biosolids management, but also to those citizens who are concerned with learning about a practical methodology for beneficial use and disposal of waste, and thus a means of improving the environment.

In line with these goals and objectives, this text is not an "engineering" text. To the contrary, it is a "user's" text. For those who are seeking a purely engineering text, the work by R. T. Haug, *Compost Engineering: Principles and Practices* (1980), is probably the best known, most widely used, and most respected on this topic.

This *Wastewater Biosolids to Compost* text is organized into fourteen chapters. At the beginning of each chapter (from Chapter 2 on), a line diagram of the ASP model composting facility/process (Figure 1.2) will be shown. The part of the composting process to be discussed and/or ancillary concerns related to the process presented in a particular chapter will be shown sequentially on the line diagram so that the reader may better understand where each topic in each chapter fits into the overall process.

Chapter 1 presents introductory information that discusses the nature of biosolids and composting in terms of definitions. Chapter 2 deals with the production of biosolids in the wastewater treatment process, including a discussion related to dewatering biosolids, and ends with the delivery of suitable feedstock (biosolids) material ready for composting. In Chapter 3 capacity and design criteria for constructing a typical ASP-type composting facility are presented. Chapter 4 discusses bulking materials used in biosolids composting. Chapter 5 describes the ASP Model Composting

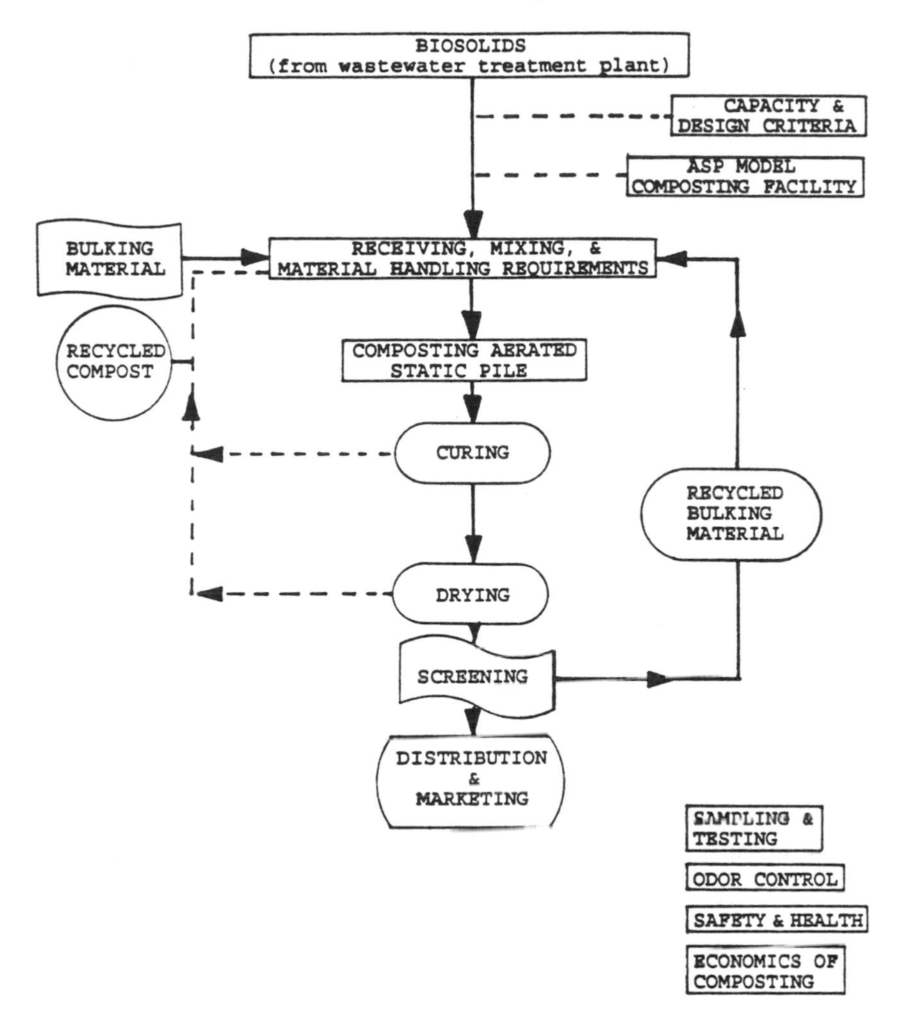

Figure 1.2 A line diagram of the ASP model aerated static pile composting process and ancillaries.

Facility. Chapter 6 deals with receiving, mixing, and material handling requirements. Chapter 7 discusses aerated static pile composting methodology. Chapter 8 covers the drying and curing process. In Chapter 9, screening and reuse of recycled bulking material and compost are covered. Chapter 10 gives an overview of the critical component of the biosolids composting process: distribution/marketing. Without a successful marketing strategy or program, the composting process is a wasted and costly enterprise. Chapter 11 discusses sampling and testing procedures required to ensure compliance with EPA's 503 rule. Chapter 12 discusses odor control. Chapter 13 focuses on the health and safety concerns related to biosolids composting. Finally, Chapter 14 discusses the economics of biosolids composting.

After reading all fourteen chapters, the reader should be able to make a rational and informed decision on whether or not benefits are to be gained from incorporating biosolids composting in his/her wastewater treatment process.

A final word before moving on to the "nuts-and-bolts" of the biosolids composting process. It is important for decision-makers to realize from the very beginning that biosolids is not a "golden goose." It will not afford the composting entity an exorbitant profit and is not problem-free (Goldstein, 1985). Even though interest in composting as a means of municipal biosolids treatment has increased dramatically since the early 1970s (Benedict et al., 1986), it entails obstacles and risks—federal regulations, equipment breakdowns, cuts in federal and state funding, public acceptance—that must be considered.

In order for biosolids composting to become a viable waste disposal alternative, public officials, engineers, operators, vendors, and citizens must work together. These key players must recognize that biosolids composting can lead to problems and unnecessary expense. This is especially the case if biosolids composting is rushed into without careful planning and much thought. All parties must know from the outset that the technology of biosolids composting is new, and therefore lacks specific objective criteria of process performance (Finstein et al., 1986).

However, biosolids composting is a growing enterprise that shows much promise for the future. While problems exist, they can and will be overcome. With 201 biosolids composting facilities on line and another 100+ under development, time and experience will help to fine-tune this important waste disposal/reuse process.

CHAPTER 2

Biosolids

INTRODUCTION

THE residual left after wastewater treatment, biosolids, is the main constituent (feedstock) of the type of compost addressed in this text. Lester (1992) and Sundstrom and Klei (1979) point out that biosolids handling and disposal constitute 25 to 40% of the total operating cost of a wastewater treatment facility. With such a large portion of operating costs spent on treatment and disposal, one might assume that biosolid treatment and disposal practices and procedures would receive a proportional amount of attention. This has not been the case, however (Lester, 1992). In order to properly design and operate a biosolids-composting facility with the goal of producing and marketing a beneficial-use end-product that is earth friendly, it is essential to pay particular attention to the product's main constituent: biosolids.

The purpose of this chapter is to present important data and information on wastewater biosolids—from the treatment process through delivery to the composting site. Moreover, the data and information will serve as foundational material for process methodology to be presented in subsequent chapters.

BIOSOLIDS: AS COMPONENTS OF THE WASTESTREAM

Domestic sewage entering the headworks of a wastewater treatment plant is composed of 99.9% water and 0.1% total solids. The total solids portion is divided about 50–50 between dissolved and suspended forms. Approximately 70% of the main raw sewage components is composed of proteins and urea (nitrogeneous compounds); sugars, cellulose,, and starches (carbohydrates); cooking oil, greases, and soaps (fats). About 30% of the solids consists of inorganic components such as metallic salts, road grit and chloride where sewerage and stormwater are combined.

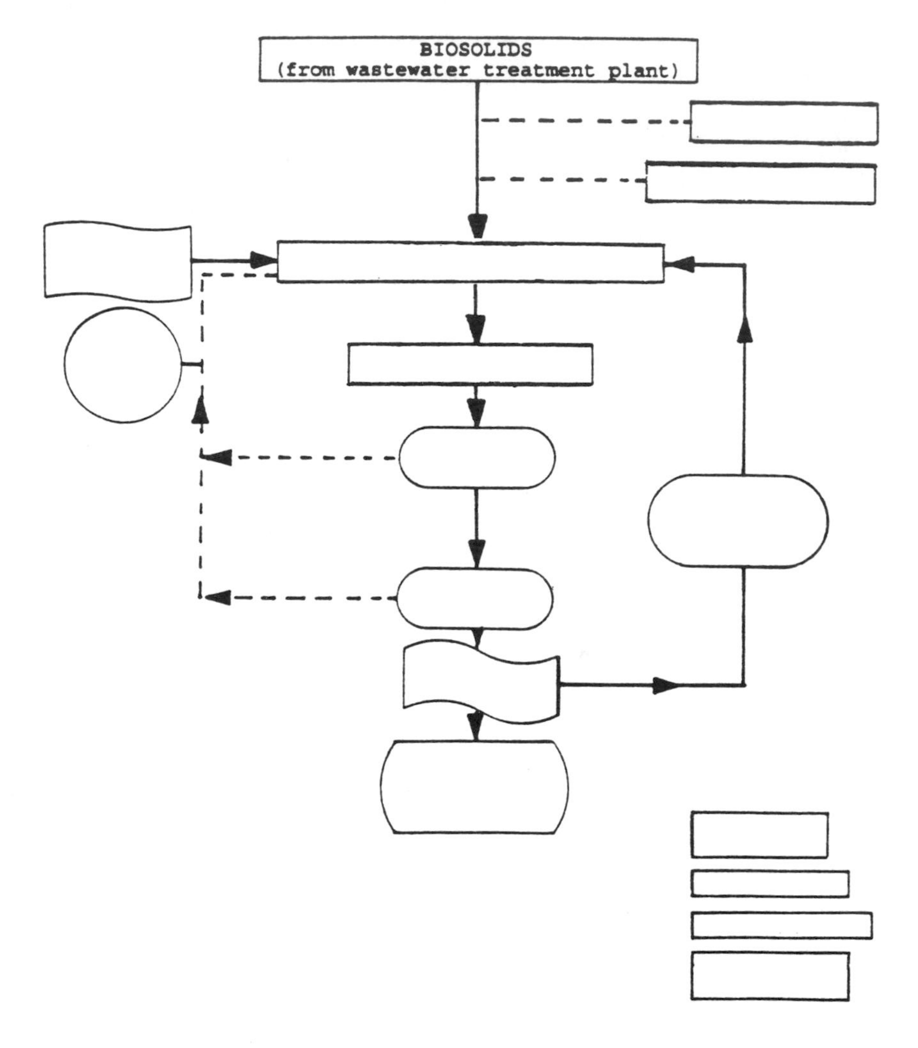

Within the wastewater treatment process, one of the main objectives from start to finish is to separate the liquid from the solid waste. While it is in the treatment process, the solids portion of the wastewater stream (biosolids) is termed in relation to its source. That is, it may be termed *primary, secondary, excess activated,* or *chemical biosolids.* It can also be named for the treatment it receives. For example, it may be called *fresh, raw, digested, elutriated* (a conditioning process), *dewatered,* or *dried.*

Biosolids are generated as a result of the wastewater treatment process. More importantly, as with the liquid portion of the wastestream, biosolids undergoes some type of treatment to facilitate its disposal. This is the case because the biosolids must be altered to a form that may be disposed or reused without endangering the environment, public health, or creating a nuisance.

Two major goals are involved in the treatment of biosolids: (1) to digest or decompose (causes a reduction in the total solids) the highly putrescible organic matter to inert or stable organic and inorganic compounds from which water will separate more readily, and (2) to reduce the volume of solids to be handled by removal of some or all of the liquid portion.

UNWANTED SOLIDS

Within the wastewater treatment process, various types or forms of solids are unwanted and, therefore, classified as waste (undesirable component parts of biosolids); thus, they are removed. The removal process is generally accomplished through screening, degritting, and skimming. Screenings, grit, skimmings, and other solids are usually small in volume, compared to other solids in the process, and are relatively easy to remove in preliminary treatment.

SCREENINGS

Screenings are materials in the influent (raw wastewater) that are caught on racks or screens with openings usually of 0.5 to 2 inches. The screens, placed at the head of the treatment plant (headworks), remove materials such as rocks, garbage, rags, and other large solids that may find their way into a collector or interceptor system and might clog pipes and damage process machinery (Masters, 1991).

GRIT

Grit can be described as small inorganic solids that are removed from the wastewater influent after screening. Sand, ashes, gravel, coffee grounds, and silt are examples of grit, all of which have settling characteristics considerably greater than organic substances (Metcalf & Eddy, 1991).

SKIMMINGS

Types of floatable material that rise in sedimentation tanks and must be skimmed from the process.

Along with clogging process piping, damaging pumps and other machinery, and as undesirable components of biosolids, screenings, grit, and skimmings also present other operational problems. Such problems include occupying space in sedimentation and digestion processes; producing odors; and interfering with the biological process that takes place in digestion.

BIOSOLIDS

Biosolids is defined as a semi-liquid waste having a total solids concentration of at least 2,500 ppm. It flows, it can be pumped, and it exhibits hindered settling characteristics in gravity settling basins. Biosolids handling and disposal includes: (1) collection of the biosolids, (2) transportation of the biosolids, (3) processing of the biosolids to a form suitable for disposal or reuse, and (4) final disposal or reuse of the biosolids.

COMPOSITION OF BIOSOLIDS

The quantity and composition of biosolids vary with the character of the sewage from which it is removed and depend upon the treatment process by which it is obtained.

In treating biosolids for composting, it is important to remove as much solids from the wastewater effluent and to produce biosolids with as high a solids concentration as possible. Cheremisinoff (1995) points out that sedimentation tank design determines the efficiency of producing highly concentrated biosolids. That is, if a new process or plant is being designed with the goal of producing biosolids of high solids concentration for eventual composting, serious consideration must be given to sedimentation tank design.

The biosolids obtained from plain sedimentation tanks is essentially the settleable solids in the raw sewage, termed *raw biosolids.* It has undergone practically no decomposition and is, therefore, highly unstable and putrescible. This type of biosolids is usually gray in color, disagreeable in appearance, and contains bits of fecal solids, garbage, sticks, and other debris. In addition, it almost always has a foul odor.

The biosolids from secondary settling tanks, following a trickling filter, consists of partially decomposed organic matter. It is usually dark brown and flocculent, more homogeneous in appearance, and has less odor than raw biosolids. The excess biosolids withdrawn from the activated biosolid process is also partially decomposed, is golden brown and flocculent, and has a rather distinct earthy odor. Biosolids from either ordinary sedimentation tanks or secondary settling tanks, with further decomposition, can become septic and cause offensive odors.

Biosolids from the chemical precipitation process is usually darker in color. The odor may be objectionable, but is not as bad as biosolids from plain sedimentation. Biosolids from the chemical precipitation process will decompose or digest, but more slowly than biosolids from other processes. The volume of biosolids produced by this process is so great that it is not practical to provide digestion facilities; therefore, other treatments are used in preparing it for disposal or reuse.

Biosolids from digestion processes has a distinct, but unoffensive odor, which varies depending on the source of the biosolids.

SOLIDS CONCENTRATION

The proportion of solids and water in liquid biosolids depends on the nature of the solids, its source—whether from primary or secondary settling tanks—and the frequency of removal from these tanks. It may vary from 1% in a watery activated biosolids to 10% or more in a concentrated raw or a digested biosolids. Concentration is important because the volume occupied is inversely proportional to the solids content. In the biosolids thickening process (to be discussed later), an approximation for determining biosolids volume reduction is well illustrated by the following equation

$$\frac{V_1}{V_2} = \frac{P_2}{P_1}$$

where

V_1, V_2 = sludge volumes before and after thickening
P_1, P_2 = percent of solid matter before and after thickening

It is desirable to handle the most concentrated biosolids possible because (1) the more highly concentrated the biosolids, the larger the amount of savings in digester space; (2) highly concentrated biosolids allows for longer digestion periods for solids; and (3) concentrated biosolids contains less water; therefore, heat requirements can be reduced in digesters.

BIOSOLIDS TREATMENT METHODS

The processes involved in biosolids treatment vary from simple gravity thickening to complete destruction by incineration. Which process is selected to accomplish the design objectives depends on one or more of the following factors:

(1) Character of the biosolids: raw, digested, or industrial

(2) Land availability
(3) Suitability of biosolids disposal by dilution
(4) Local possibilities for using biosolids as a soil conditioner or fertilizer
(5) Climate
(6) Capital and operating costs
(7) Size and type of wastewater treatment plant
(8) Proximity of the plant to residual areas and local air pollution control regulations

This text includes a range of methods used at treatment plants to the point of suitability for delivery to the composting facility. In preparing a biosolids product suitable for composting, keep in mind that all biosolids treatment methodologies have three main objectives: (1) under current regulations (EPA's 503 requirements) biosolids must be treated in such a manner that pathogen and vector attraction reduction is accomplished, (2) the biosolids to be delivered to the composting facility must be dewatered to at least 40%, and (3) the treated biosolids must be of suitable quality to enhance the composting process.

The biosolids treatment methods to be discussed in this text include the following:

(1) Thickening
(2) Stabilization
(3) Conditioning
(4) Dewatering
(5) Volume reduction
(6) Disposal

Along with the biosolids treatment methods, the biosolids unit processing alternatives addressed in this text are outlined in Figure 2.1.

BIOSOLIDS THICKENING

Biosolids thickening is a physical process that is accomplished by gravity (solids allowed to settle to the bottom), flotation (solids are floated to the top), or centrifugation. Gravity thickening is generally used on biosolids from untreated primary processes. Dissolved-air flotation and centrifugation, on the other hand, are used to thicken waste-activated biosolids (Metcalf & Eddy, 1991).

The primary objective of biosolids thickening processes is to remove as much water as possible before the biosolids moves on to other treatment processes (Davis & Cornwell, 1991). Keeping in mind that biosolids thickening is accomplished in wastewater treatment to increase the efficiency of

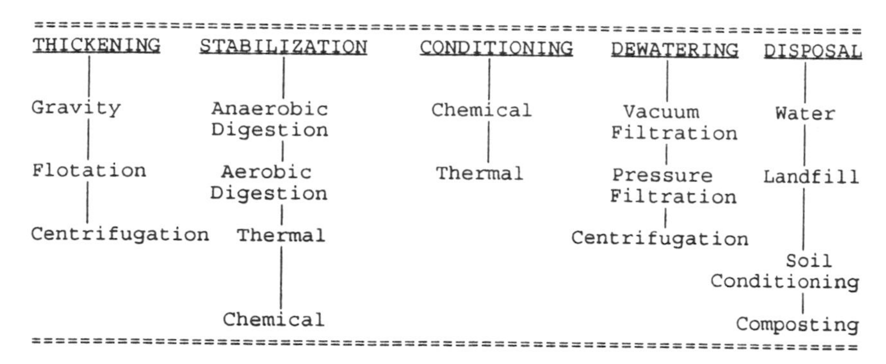

Figure 2.1 Biosolids unit processing alternatives.

the treatment processes to follow, Vesilind (1980) defines the biosolids thickening process as the concentration of solids to less than 15% solids.

Gravity Thickening

Similar in structure to circular sedimentation basins, gravity thickeners process a thin concentration of biosolids to a more dense biosolids. The gravity thickener's use is largely restricted to the watery excess biosolids from the activated biosolids process, especially in large plants where the biosolids is sent directly to digesters instead of the primary tanks. It may also be used to concentrate biosolids from primary tanks or a mixture of primary and excess biosolids prior to high rate digestion.

The gravity thickening tank is generally equipped with slowly moving biosolids scrappers that have a vertical, picket fence-like structure attached, which serves to agitate the biosolids and dislodge entrapped liquid and gas bubbles (McGhee, 1991). Biosolids is usually pumped continuously from the settling tank to the thickener, which has a low overflow rate, so that the excess water overflows and the sludge biosolids solids concentrate on the bottom. A blanket of biosolids is maintained by controlled removal, which may be continuous at a low rate. A biosolids with a solids content of 10% or more can be produced by this method. This means that with an original biosolids of 2%, about four-fifths of the water has been removed; thus, one of the objectives in biosolids treatment has been accomplished.

Flotation Thickening

The popularity of flotation thickening units has increased at sewage treatment plants, especially for handling waste-activated biosolids. With activated biosolids, these units have the advantage over gravity-thickening tanks of offering higher solids concentrations and lower initial capital expenditure (Corbitt, 1990).

The objective of flotation thickening is to attach a minute air bubble to suspended solids causing them to separate from the water in an upward direction. This separation is due to the fact that the solid particles have a gravity that is lower than water when the bubble is attached. Moreover, when activated biosolids are thickened in this process, concentration at the surface is facilitated by the close density relationship between the water and the activated biosolids, which allows them to be readily buoyed to the surface (McGhee, 1991).

Dissolved air flotation depends on the formation of small-diameter bubbles resulting from air released from solution after being pressurized to 40 to 60 psi. Since the solubility of air increases with pressure, substantial quantities of air can be dissolved. In current flotation practice, two general approaches to pressurization are used: (1) air charging and pressurization of recycled clarified effluent or some other flow used for dilution with subsequent addition to the feed biosolids; and (2) air charging and pressurization of the combined dilution liquid and feed biosolids.

Air in excess of the decreased solubility, resulting from the release of the pressurized flow into a chamber at near atmospheric pressure, comes out of solution to form the minute air bubbles. Biosolids solids are floated by the air bubbles that attach themselves to and are enmeshed in the floc particles. The degree of adhesion depends on the surface properties of the solids. When released into the separation area of the thickening tank, the buoyed solids rise under hindered conditions analogous to those in gravity settling and can be called *hindered separation* or *flotation*. The upward moving particles form a sludge blanket on the surface of the flotation thickener.

Several variables affect the operation of flotation thickening systems. Obviously, the type and quality of the biosolids affect the unit's performance. Other important parameters include pressure, feed solids concentration, recycle ratio, detention time, air-to-solids ratio, solids and hydraulic loading rates, and chemicals used.

Flotation thickening is most applicable to activated biosolids, but higher float concentrations can be achieved by combining primary with activated biosolids. Equal or greater concentrations may be achieved by combining biosolids in gravity thickening units.

Centrifugation

Although primarily used in dewatering, centrifugation has demonstrated the capability of thickening a variety of biosolids, but is generally limited to thickening waste-activated biosolids (Metcalf & Eddy, 1991). Centrifuges are compact, simple, flexible, self-contained units, and the capital cost is relatively low. However, disadvantages include high maintenance and power costs and a poor solids-capture efficiency if chemicals are not used. *Note:* The dewatering section of this chapter contains a more detailed discussion of centrifuges.

BIOSOLIDS STABILIZATION

If biosolids are to be land-applied, stabilization of biosolids in wastewater treatment is important because of the need to reduce the volume of the thickened biosolids still further, to eliminate offensive odors, to reduce the possibility of putrefaction, and to render the remaining solids relatively pathogen-free (Peavy et al., 1991). Thus, the stabilization of biosolids to be used in the composting process is important. Biosolids stabilization can be accomplised by aerobic digestion, anaerobic digestion, and by thermal and chemical stabilization.

Anaerobic Digestion

In anaerobic digestion, biosolid digestion is carried out in the absence of free oxygen by anaerobic organisms. Therefore, what is really occurring during this process is anaerobic decomposition. As pointed out earlier, the solid matter in raw biosolids is about 70% organic and 30% inorganic. Much of the water in wastewater biosolids is "bound" water, which will not separate from the biosolids solids. The facultative and anaerobic organisms break down the complex molecular structure of these solids, thereby setting free the "bound" water and obtaining oxygen and food for their growth.

Anaerobic digestion involves many complex biochemical reactions and is dependent on many interrelated physical and chemical factors. The anaerobic digestion process is complex, but for purposes of simplification it can be summarized in two steps: (1) conversion of organic materials to volatile acids and (2) conversions of volatile acids into methane (Haller, 1995). One way to visualize this two-step operation is to think of an assembly line in a factory where the first line receives the raw materials and fashions them into another form, which is further refined on a second assembly line into the final product.

In the first step (waste conversion), acid-forming bacteria attack the soluble or dissolved complex solids, such as fats, proteins, and carbohydrates. From these reactions organic acids, at times up to several thousand ppm, and gases, such as carbon dioxide and hydrogen sulfide, are formed. This is known as the *stage of acid fermentation* and proceeds rapidly. It is followed by a period of acid digestion in which the organic acids and nitrogenous compounds are attacked by bacteria called *acid formers* and liquified at a much slower rate (Masters, 1991).

In the second stage of digestion, known as the *period of intensive digestion, stabilization and gasification,* the more resistent nitrogenous materials, such as the proteins, amino acids and others, are attacked by microorganisms. This stage of the digestion process is sensitive and the pH value must be maintained from 6.8 to 7.4. Large volumes of gases with a 65 or higher percentage of methane (CH_4) are produced. Methane is an odorless, highly flammable gas that can be used as a fuel. The organisms (bacteria), which convert organic acids to methane and carbon dioxide gases, are called *methane formers.* The solids remaining (the constituents this text is concerned with) are relatively stable or only slowly putrescible, can be disposed of without creating objectionable conditions, and have value in agriculture via liquid land application and compost.

In a healthy, well-operating digester, both of the above steps are taking place continuously and at the same time. Fresh wastewater solids are being added at frequent intervals, with the stabilized solids being removed for further treatment or disposal at less frequent intervals. The supernatant digester liquor, the product of liquefaction and mechanical separation, is removed frequently to make room for the fresh solids and the gas is, of course, being removed continuously.

While all stages of digestion may be proceeding in a digester at the same time, with the acids produced in the first stage being neutralized by the ammonia produced in subsequent stages, the best and quickest results are obtained when an overall pH of 6.8 to 7.4 predominates. The first stage of acid formation would be evident only in starting up digestion units. Thus, once good alkaline digestion is established, the acid stage is not apparent unless the normal digestion becomes upset by overloading, or chemicals, or for other reasons. It is critical to the overall process to maintain balanced populations of acid formers and methane formers. The methane formers are more sensitive to environmental conditions and slower growing than the acid-forming group of bacteria and control the overall reactions.

The process of digestion can be measured by the destruction of organic matter (volatile solids), by the volume and composition of gases produced, and by the pH, volatile acids, and alkalinity concentration. No one parameter or test can be used to predict problems or to control a digester. Thus, several parameters should be considered together when troubleshooting problems or when operating digesters.

For the purposes of this text, the discussion of anaerobic digestion will end here by simply pointing out that anaerobic digestion of wastewater biosolids contributes significantly to the goal of this text; that is, to the beneficial production of a wastewater biosolids by-product (Metcalf & Eddy, 1991).

Aerobic Digestion

Aerobic digestion is an extension of the activated biosolids aeration process, whereby waste primary and secondary biosolids are continually aerated for long periods of time. In aerobic digestion, the microorganisms extend into the endogenous respiration phase, where materials previously stored by the cell are oxidized, resulting in a reduction in the biologically degradable organic matter. In simple terms, the endogenous stage can be defined as the point in which food is depleted to such a degree that the microorganisms begin to consume their own protoplasm through oxidation. This organic matter from biosolids cells is oxidized to carbon dioxide, water, and ammonia. The ammonia is further converted to nitrates as the digestion process proceeds.

Eventually, the oxygen uptake rate levels off and the biosolids matter is reduced to inorganic matter and relatively stable volatile solids.

The major advantage of aerobic digestion is that it produces a biologically stable end product suitable for subsequent treatment in a variety of processes. Volatile solids reductions similar to anaerobic digestion are possible.

Aerobic digestion is affected by several factors: (1) biosolids temperature, (2) rate of biosolids oxidation, (3) biosolids loading rate, (4) system oxygen requirements, (5) biosolids age, and (6) biosolids solids characteristics.

As with almost all treatment processes, aerobic digestion entails both advantages and disadvantages. The advantages most often identified include:

(1) A humuslike, biologically stable end product is produced.

(2) The stable end product has no odors; therefore, simple land disposal, such as in lagoons, is feasible.

(3) Capital costs for an aerobic system are low when compared with anaerobic digestion and other processes.

(4) Aerobically digested sludge usually has good dewatering characteristics. When applied to sand drying beds, it drains well and redries quickly if rained on.

(5) The volatile solids reduction can be equal to those achieved by anaerobic digestion.

(6) Supernatant liquors from aerobic digestion have a lower BOD (generally lower than 100 ppm) than those from anaerobic digestion.

(7) There are fewer operational problems with aerobic digestion than with the more complex anaerobic form because the system is more stable. As a result, less skilled labor can be used to operate the facility.

(8) Compared with anaerobic digestion, more of the biosolids' basic fertilizer values are recovered (Metcalf & Eddy, 1991).

The major disadvantage associated with aerobic digestion is high power costs. Unlike anaerobic digestion, aerobic digestion requires the supply of oxygen, which is energy consumptive (Peavy et al., 1985). At small wastewater treatment plants, the power costs may not be significant but they might be at larger plants. Experience suggests another disadvantage in that aerobically digested biosolids does not always settle well in subsequent thickening processes. This situation leads to a thickening tank decant having a high solids concentration. Two other disadvantages associated with aerobically digested biosolids are (1) biosolids does not dewater easily by vacuum filtration and (2) the variable solids reduction efficiency changes with varying temperature (Metcalf & Eddy, 1991).

Thermal Stabilization

Thermal stabilization is a heat process by which the bound water (water associated with biosolids) of the biosolids solids is released by heating the biosolids for short periods of time. Although listed here as a stabilization process, Metcalf & Eddy (1991) point out that thermal treatment can be used both for stabilization and for conditioning of biosolids.

Exposing the biosolids to heat and pressure coagulates the solids, breaks down the cell structure, and reduces the hydration and hydrophilic (water loving) nature of the solids. The liquid portion of the biosolids can then be separated by decanting and pressing.

Chemical Stabilization

Chemical stabilization is a process whereby the biosolids medium is treated with chemicals in different ways to stabilize the biosolids solids. Because the cost of chemicals can be high (in some cases prohibitive), chemical stabilization processes are not broadly used in the wastewater industry (Corbitt, 1990).

Chlorine Stabilization

Stabilization reaction by chlorine is almost instantaneous with very little volatile solids reduction in the biosolids. There is some breakdown of organic material and formation of carbon dioxide and nitrogen. Application of high doses (normally in the range from about 650 to almost 5,000 mg/L, depending on the type of biosolids and solids concentration) of chlorine gas directly to the biosolids in an enclosed reactor produces a stabilized biosolids.

The stabilized biosolids will have a pH of 2.5 to 4.5 and chlorine residual of 200 to 400 mg/L. The stabilized biosolids will have chlorine smell and light brown color. Total solids, suspended solids, and volatile solids concentrations will be about the same as the raw biosolids. Because the chlorine-treated biosolids has a low pH value it may require adjustment before conditioning. In addition, chlorine-treated biosolids may contain undissolved heavy metals and chlorinated compounds that limit its suitability for land-application activities; thus, chlorine stabilization is rarely used as a biosolids stabilization process.

Lime Stabilization

Lime stabilization has a long history of use for stabilization of biosolids. For example, lime stabilization processes may be used to treat raw primary, waste-activated, septage and anaerobically digested biosolids. Lime works well for biosolids stabilization because it increases the pH high enough to destroy most microorganisms and limit odor production. Additionally, Vesilind (1980) points out that "lime is good for sludge (biosolids) stabilization because nothing is oxidized and no dangerous substances are formed" (p. 75).

The process involves mixing a sufficiently large quantity of lime with the biosolids to increase the pH of the mixture to 12 or more. Vesilind (1980) points out that the pH must be maintained between 12.2 and 12.4 to ensure that the biosolids is stabilized and the pathogens are destroyed.

Along with reducing bacterial hazards and odor to a negligible value, lime stabilization also improves vacuum filter performance, provides a satisfactory means of stabilizing biosolids prior to ultimate disposal. Also by maintaining the pH above 12 for two hours or more, the total reduction in microorganisms will be substantially higher than those obtained in digestion processes (McGhee, 1991).

Although lime is available in a number of forms, the most commonly used forms for biosolids stabilization are quicklime and hydrated lime. Quicklime (unslaked lime) consists almost entirely of calcium oxide, CaO. Quicklime does not react uniformly when applied directly to sludge, but must first be converted to the hydrated form, $Ca(OH)_2$. Hydrated or slaked lime is a powder obtained by adding sufficient water to quicklime to satisfy its affinity for water.

BIOSOLIDS CONDITIONING

Biosolids conditioning is a process whereby biosolids solids are treated with chemicals or various other means to improve production rate, cake solids content, and solids capture, all of which prepare the biosolids for dewatering processes. Several biosolids conditioning processes have been used in wastewater treatment. In this text, the two most commonly used methods, addition of chemicals and heat treatment, will be discussed.

Chemical Conditioning

Chemical conditioning (biosolids conditioning) prepares the biosolids for better and more economical treatment with dewatering equipment. This is accomplished by reducing the biosolids moisture content to 60 to 85%. Many chemicals have been used such as organic polymers, alum, ferrous sulfate, and ferric chloride with or without lime, and others. All of them are more easily applied in liquid form (Metcalf & Eddy, 1991). The choice of chemical type depends on the nature of the biosolids to be conditioned, the major determining factor usually being local cost. In recent years, ferric chloride and organic polymers have become increasingly popular (Davis & Cornwell, 1991).

The addition of the chemical to the biosolids lowers or raises its pH value to a point where small particles coagulate into larger ones, and the water in the biosolids solids is given up more readily. No one pH value is best for all biosolids. Different biosolids such as primary, secondary, and digested biosolids, and different biosolids of the same type, have different optimum pH values, which must be determined for each biosolids by conducting pilot studies or by trial and error.

Thermal Conditioning

Thermal conditioning destroys the biological cells in biosolids which "permits a degree of moisture release not achieved in other conditioning processes" (McGhee, 1991, p. 497). There are several basic processes for thermal treatment of biosolids. In this text two processes are covered: wet air oxidation and heat treatment.

Wet air oxidation is the flameless oxidation of biosolids at temperatures of 450 to 550°F and pressures of about 1,200 psig. The heat treatment type is similar to the wet air oxidation type, but is carried out at temperatures of about 350 to 400°F and pressures of 150 to 300 psig. These two processes work much like a pressure cooker; that is, water that is bound up in the biosolids is released, which in turn facilitates the dewatering process (Davis & Cornwell, 1991). Wet air oxidation reduces the biosolids to an ash and heat treatment improves the dewaterability of the biosolids. The lower temperature and pressure heat treatment is more widely used than the oxidation process.

When the biosolids is heated, heat causes water to escape. Thermal treatment systems release water that is bound within the cell structure of the biosolids and thereby improves its dewatering and thickening characteristics. The oxidation process further reduces the biosolids to ash by wet incineration (oxidation). Biosolids is ground to a controlled particle size and pumped to a pressure of about 300 psi. Compressed air is added to the biosolids (wet air oxidation only); the mixture is brought to a temperature of about 350°F by heat exchange with treated biosolids and direct stream injection, and is then processed (cooked) in the reactor at the desired temperature and pressure. The hot treated biosolids is cooled by heat exchange with the incoming biosolids. The treated biosolids is settled from the supernatant before the dewatering step.

The same basic process is used for wet air oxidation of biosolids by operating at higher temperatures (450 to 640°F) and higher pressures (1,200 to 1,600 psig). The wet air oxidation process is based on the fact that any substance capable of burning can be oxidized in the presence of water at temperatures between 250°F and 700°F. Wet air oxidation does not require preliminary dewatering or drying as do conventional air combustion processes. However, the oxidized ash must be separated from the water by vacuum filtration, centrifugation, or some other solids separation technique.

An advantage of thermal treatment is that a more readily dewaterable biosolids is produced than with chemical conditioning. Dewatered biosolids of 30 to 40% have been achieved with heat treated biosolids at relatively high loading rates on dewatering equipment, which is about three times the rates with chemical conditioning. The process also provides effective disinfection of the biosolids, all of which produces a biosolids conditioned for easier dewatering.

Heat treatment processes also have disadvantages, however. For example, heat treatment ruptures the cell walls of microbial organisms, thereby releasing not only the water but also some bound organic material. This returns to the solution some organic material previously converted to particulate form and creates other fine particulate matter. The breakdown of the biological cells as a result of heat treatment converts these previously particulate cells back to water and fine solids. While this aids the dewatering process, it creates a separate problem of treating this highly polluted sidestream of liquid from the cells.

Because of this disadvantage and others, such as high initial capital costs, higher levels of competence and training required of operating personnel, and the significant production of odorous gases, the construction of new heat treatment facilities has slowed considerably in the United States (Metcalf & Eddy, 1991).

DEWATERING

After pretreatment of biosolids by the conditioning process, the next step in the biosolids unit process is dewatering. Exactly what are the objectives of biosolids dewatering? According to the U.S. EPA (1982), the objectives of dewatering "are to remove water and thereby reduce the [biosolids] volume, to produce a [biosolids] which behaves as a solid and not a liquid, and to reduce the cost of subsequent treatment and disposal processes" (p. 2).

Figure 2.2 shows a dewatered biosolids sampling point and a dewatered biosolids sample. The biosolids sample is on a conveyor belt leading to the intake of a multiple hearth incinerator. This particular biosolids sample is not the dewatered cake product that is normally used in biosolids-derived composting.

Figure 2.2 Dewatered biosolids sampling point and dewatered biosolids sample.

According to Shimp et al. (1995), dewatering is very important because "it profoundly impacts the economics, functioning, and required capacity of downstream unit operations" (pp. 44–49). With regard to the downstream unit operation that produces a beneficial-use end product, such as a compost, dewatering is very critical.

This is especially the case in preparing biosolids for transportation and use in composting, where a dewatered cake of from 18 to 25% solids is desired. Transporting biosolids is an expensive undertaking for many facilities. As a case in point, consider the expense involved with transporting biosolids over long distances, which is not unusual. For example, Padmanabha et al. (1994) report about one plant where more than 1,800 wet ton/day of cake is produced and then hauled approximately 50 miles to the land-application site. In this instance, it would be even more costly to transport the biosolids product if it had a higher water than solids content.

Moreover, higher solids content is important because it reduces need for space, fuel, labor, equipment, and size of the composting facility (Epstein & Alpert, 1984). Dewatering methods also affect the amount of bulking agent needed for composting. Thus, the performance of solids dewatering methods used in preparing biosolids for disposal as a legal and marketable beneficial-use end product is very important.

Biosolids dewatering can be accomplished in a variety of ways. In this text, vacuum filtration, pressure filtration, centrifuging, and selected other methods will be discussed.

Vacuum Filtration

Metcalf & Eddy (1991) point out that in recent years the vacuum filtration method of dewatering biosolids has declined because of high operating and maintenance costs as well as advances in technology. However, since several facilities still use this method of dewatering biosolids, it will be covered here.

The vacuum filter for dewatering biosolids consists of a drum over which is laid the filtering medium consisting of a cloth of cotton, wool, nylon, fiber glass or a plastic or stainless steel mesh, or a double layer of stainless steel coil springs. The drum with horizontal axis is set in a tank with about one quarter of the drum submerged in conditioned biosolids. Valves and piping are so arranged that, as a portion of the drum rotates slowly in the biosolids, a vacuum is applied on the inner side of the filter medium, drawing out water from the biosolids and holding the biosolids against it. Application of the vacuum is continued as the drum rotates out of the biosolids and into the atmosphere. This pulls water away from the biosolids, leaving a moist mat or cake on the outer surface. This mat is scraped, blown, or lifted away from the drum just before it enters the biosolids tank again.

The common measure of performance of vacuum filters is the rate in pounds per hour of dry solids filtered per square foot of filter surface. This rate may vary from a low of 2.5 for activated biosolids to a high of 6 to 11 for the best digested primary biosolids. The moisture content in the biosolid cake also varies with the type of biosolids from 80 to 84% for raw activated biosolids to 60 to 68% for well-digested primary biosolids.

Pressure Filtration

Pressure filtration is usually accomplished by using belt filter or recessed plate filter presses. In the use of either method, filtration is accomplished in a fashion similar to vacuum filtration, where biosolids are separated from the liquid. At present, recessed plate filter press units are the most common type used. However, use of belt filter presses for dewatering municipal wastewater biosolids has increased (see Figure 2.3).

Like vacuum filtration, a porous medium is used in recessed plate filter presses to separate solids from the liquid. The solids are captured in the media pores; they build up on the media surface; and they reinforce the media in its solid-liquid separation action. Biosolids pumps provide the energy to force the water through the media.

One of the major advantages of using pressure filtration to dewater biosolids is that it can be accomplished quickly in a small space. Moreover, Vesilind (1980) reports that filter presses have performed well (pro-

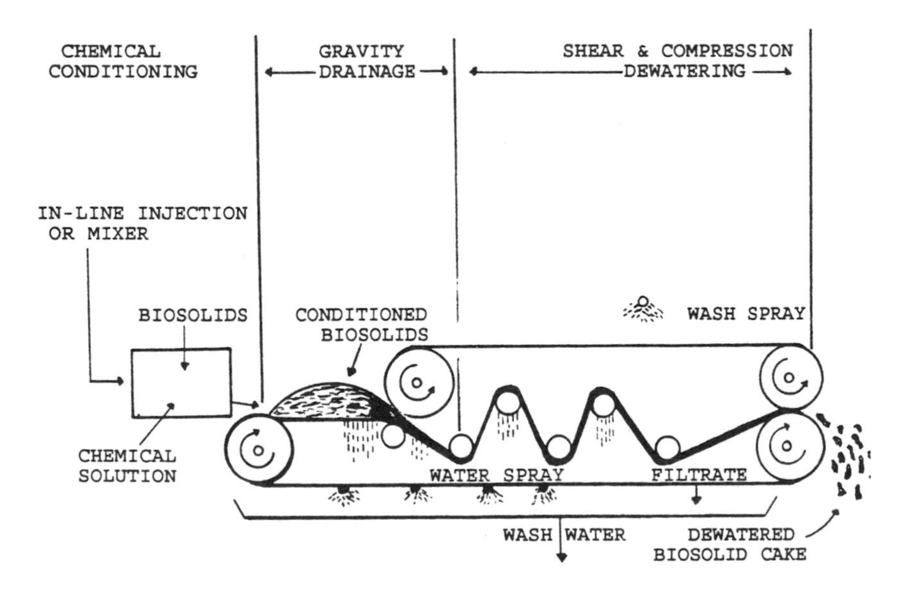

Figure 2.3 A belt filter press used for dewatering biosolids.

ducing >30% cake solids in some cases), compared to the vacuum filter (in come cases with only around 18% cake solids produced). However, compared to other dewatering methods, it has major disadvantages of (1) requiring batch operation, and (2) high operation and maintenance costs.

Centrifugation

According to Haller (1995), centrifugation used in dewatering biosolids, compared to the other methods previously discussed, has the advantage of producing a cake that is more consistent and less likely to generate odors. Centrifugation has the additional advantage of requiring less space than other types of dewatering equipment. One disadvantage that is commonly identified with centrifuges is their higher power costs. However, these higher operating costs may be offset by the lower initial costs (Metcalf & Eddy, 1991).

In centrifugation, solids are separated from liquid through sedimentation and centrifugal force. In a typical unit, biosolids is fed through a stationary feed tube along the centerline of the bowl through a hub of the screw conveyor, Mounted inside the rotating conical bowl, the screw conveyor rotates at a slightly lower speed than the bowl. Biosolids leaves the end of the feed tube, is accelerated, passes through the ports in the conveyor shaft, and is distributed to the periphery of the bowl. In the process, solids settle through the liquid pool, are compacted by centrifugal force against the walls of the bowl, and are finally conveyed by the screw conveyor to the drying or beach area of the bowl. The beach area is an inclined section of the bowl where further dewatering occurs before the solids are discharged. Separated liquid is discharged continuously over adjustable weirs at the opposite end of the bowl (see Figure 2.4).

In the past, wastewater treatment managers have been reluctant to select centrifugation over other dewatering processes. Usually, a functional analysis is conducted to evaluate the best alternative dewatering process to use. The two alternatives commonly evaluated are the belt filter press and the centrifuge for dewatering.

Such analysis generally shows an advantage for the belt filter press system compared to centrifugation in capital costs, power requirements, and operating expense. In the past, centrifugation demonstrated only one major advantage: odor control. This was usually the case because the centrifugation unit process was totally enclosed, allowing the off-gases to be vented directly to an odor control system.

A recently documented study compared the operating costs for a belt filter press and centrifuge (Shimp et al., 1995). The results indicated that despite the concern often expressed about the power costs associated with centrifuge dewatering, "power is typically the smallest component of

Figure 2.4 State-of-the-art high-solids centrifuge used for dewatering biosolids.

dewatering operations and maintenance costs" (pp. 44–49). The major cost components involved with dewatering are usually associated with chemical conditioning and operating labor.

In recent years, centrifugation has been making a comeback in dewatering biosolids, because plants have become more concerned with odor control. Later in this text a detailed account of odor control and odor control problems will be presented. For now, suffice it to point out that odor and its control must be addressed not only at the wastewater treatment plant but also at the composting site. How serious is the problem with odor generation? According to Goldstein (1985), "the number one enemy of a sewage [biosolids] composting facility is odor" (p. 20).

In addition to reducing odors, recent centrifuge designs have reduced maintenance problems and costs and have improved centrifuged cake solids to the point where they are comparable to those produced in other dewatering processes. In some cases centrifugation of biosolids is producing a biosolids cake product so "dry" that it leads to other problems. As a case in point, consider the dilemma faced by Finley and Morse (1996) recently.

At HRSD's Peninsula Composting Facility in Newport News, Virginia, dewatered biosolids is delivered to the composting facility in Ram-E-Ject

Trucks. Prior to the installation of high solids centrifuges at the wastewater treatment facilities that provide biosolids feedstock to the composting facility, unloading the biosolids cake was never a problem. After installation of the centrifuges, however, the trucks were delivering a biosolids cake product that was 30% solids or greater. This is where the problem started. The trucks had difficulty depositing their "dry" biosolids cake product onto the mixing pad because the cake was so "dry" that it would not readily move out of the trailer.

To solve this problem, Finley and Morse devised a methodology that, in the early stages, seems to be working. They mixed a portion of bentonite clay (an inert substance) with water that produced a slurry that, when used to coat the sides and bottom of the truck trailer, allowed the dry biosolids to slide freely from the trailer and onto the mixing pad.

Another reason for the recent return to using centrifugation of biosolids for dewatering is the development of high G-force centrifuges. Operated at up to 4,000 G, these high G-force centrifuges produce solids in the 32 to 35% range. Along with increase in cake dryness, new improved materials, including the use of stainless steel for wet parts, and proprietary changes have improved performance and limited downtime. Studies have shown that the energy requirements to operate a high G-force centrifuge is site-specific and is influenced by the centrifuge operating schedule (M'Coy et al., 1994).

Other Methods of Dewatering Biosolids

The ongoing quest to develop inexpensive and efficient biosolids dewatering methods is being conducted worldwide. In the past ten years, for example, developments in biosolids dewatering have included solar and electroacoustic methods. These two dewatering methods and several others under development may save in operating space requirements and in costs by, for example, decreasing the amount of bulking agent used in composting by increasing the solids content to as much as 35%. More research is needed in this area, but the current trend toward developing more efficient and cost-effective dewatering methodologies is encouraging.

With Composting as the End-Use Methodology, Which Dewatering Method Is Best?

Thus far, a variety of mechanical dewatering processes have been described. Dewatering processes such as sand drying beds and biosolids lagoons have not been discussed because they are not normally used in the preparation of biosolids for composting. The three major dewatering methods that have been discussed include vacuum filtration, pressure filtration (both belt pressure and pressure plate filtration), and centrifugation.

When determining which type of dewatering process is best suited for a particluar facility, several factors come into play. In the first place, the decision-maker(s) in an operational wastewater facility must take a look at the current dewatering system. Obviously, if a plant's dewatering process is already installed, is fully operational, and is producing a biosolid product with 20% solids, it might not be feasible or cost effective to replace this system. On the other hand, if the presently installed dewatering process is scheduled for major retrofit or replacement and if the end-use goal is production of a product suitable for composting, then the decision-maker should compare available dewatering processes and decide which one is best suited for a particular operation.

A second factor to be considered is the cost-benefit from such an installation. Along with the cost-benefit, there is another important consideration to take into account: availability of space. It does little good to choose a process for installation in an existing facility where space is limited or unavailable. To acquire a better understanding of the space requirements for installation of centrifuges, refer to Figures 2.5 and 2.6.

As stated previously, in order to determine which dewatering process type is best for replacing a present system or for installing in a new facility, it is prudent to make a comparison between the process units that are available. This comparison can best be done by looking at the advantages and disadvantages of the two types of dewatering processes that would most likely be used in the dewatering of municipal wastewater biosolids for use in composting. The U.S. EPA (1982) points out the two systems most commonly selected for this purpose are the belt filter press and the solid bowl centrifuge.

In order to demonstrate the advantages and disadvantages of the two most commonly used biosolids dewatering processes, important comparisons are provided in Tables 2.1 and 2.2. *Note:* Every system has its own inherent advantages and disadvantages when compared to other systems. However, many of the disadvantages listed in Table 2.2 have been corrected or are undergoing factory-made adjustments to refine each process to enhance operating performance.

To make a determination as to which dewatering process to purchase to replace existing dewatering equipment or to install in a new construction project, the comparisons provided in Tables 2.1 and 2.2 may be helpful. However, several other considerations must be taken into account before making the final selection. As a case in point, it is beneficial to consider the following typical case study, which details such a decision-making process. For thc purpose of this case study, the wastewater treatment plant will be called the Alpha Plant.

Figure 2.5 A typical centrifuge installation.

Figure 2.6 Associated piping for centrifuge centrate installed on the underside of the installation shown in Figure 2.5. This illustration points out the space required for an entire centrifugation installation.

TABLE 2.1. **Dewatering Processes: Comparison of Advantages.**

Solid Bowl Centrifuge	Belt Filter Presses
Little or no odor problems with fast startup and shutdown capabilities	Capable of producing drier cake than most other processes
Requires a relatively small space for installation	Low power requirements
Does not require continuous operator attention	Low noise and vibration
Can operate with a highly variable feed solids concentration on many biosolids types	Process is open to view; thus, it is easier for operator to see and understand the operation
Can be used either for dewatering or thickening	Continuous operation
Fewer units are needed since high rates of feed per unit	
Smaller doses of chemical treatment required	

Source: Adapted from EPA's *Dewatering Municipal Wastewater Sludges,* EPA-625/1-82-014, Oct. 1982.

TABLE 2.2. **Dewatering Processes: Comparison of Disadvantages.**

Solid Bowl Centrifuge	Belt Filter Presses
Internal components can be high-maintenance items	Typically require greater polymer dosages than centrifuges
Prescreening or grinding in the feed stream may be necessary	Very sensitive to incoming feed characteristics and chemical conditioning
Requires highly skilled maintenance personnel in plants where scroll maintenance is performed	Machines hydraulically limited in throughput
Noise is very noticeable	Wash water requirement for belt spraying can be significant
Vibration must be accounted for in designing ancillary support equipment	Required frequent washdown of area around press
High power consumption	Require prescreening or grinding of biosolids to remove large objects and fibrous material
Since the process is fully enclosed, flow conditions can be overlooked	Emits noticeable odors if the biosolids is poorly stabilized
	Required great deal of operator attention

Source: Adapted from EPA's *Dewatering Municipal Wastewater Sludges,* EPA-625/1-82-014, Oct. 1982.

CASE STUDY

The Alpha Plant had utilized pressure filter presses for dewatering biosolids since 1979, adding lime and ferric chloride for conditioning. Past experience demonstrated that the presssure filter presses and all associated equipment required a large amount of maintenance and had been unreliable.

The Alpha Plant, like many other wastewater treatment plants, is a total quality management (TQM) work center where empowerment of employees had been effected (i.e., employees are encouraged to speak up and contribute their ideas for increasing plant efficiency, productivity, and improving employee well-being). During one of the Alpha Plant's monthly TQM meetings, the solids handling chief operator suggested that since improvements in centrifuging biosolids had been made in recent years, it might be a good idea to replace the present dewatering system, which was due for extensive retrofitting and repair, with centrifuges. Using the total quality management team method of decision-making, the "Team" agreed to contact a centrifuge supplier and to arrange for a pilot test at the Alpha Plant.

A centrifuge manufacturer and supplier agreed to install a centrifuge pilot study at the Alpha Plant. After installation, plant personnel assisted the manufacturer's representatives in conducting the pilot study. The results were found to be better than expected.

After determining that a changeover to the centrifuge process would result in a better biosolids product, the Team proceeded to determine the economic feasibility of replacing the pressure filter press system with a centrifuge system. *Note:* The reader should remember that it is the economic bottom-line (cost) that drives all decisions such as the one that the Team was making.

To justify replacement and convince the district manager for utilities, the Alpha Plant manager documented his findings in the following report.

ALPHA WWTP CENTRIFUGE PILOT STUDY

PROCEDURE

Prior to the arrival of the centrifuge pilot prototype, the Alpha Plant determined that to match the dewatering results of the pressure filter process it was necessary to produce 25% total solids from the centrifuge. Test results indicated an actual dewatered product with 27% total solids.

The criteria examined during the pilot study was a comparison between a centrifuge facility and the plant's pressure filter press system for the following:

(1) Electrical power costs
(2) Major repairs and replacement costs
(3) Chemical costs
(4) Hauling/landfill costs
(5) Personnel

With the results of the operation and maintenance comparison already known, the study looked at the possibilities of the installation configuration as well as looking at other facilities. The following criteria were listed:

(1) Minimize pumping requirements as well as number of pumps necessary.
(2) Minimize handling of the dewatered solids as well as attempting to use flat belt conveyors if at all possible.
(3) Minimize the downtime of the facility during construction.

To aid in determining the best way to meet these criteria, the assistance of the Alpha Plant engineering department, the plant's solid handling maintenance operators, chief operator, and plant superintendent was enlisted.

The next step was to determine the equivalent uniform annual cost for each system. After the centrifuge pilot plant testing phase, an operation and maintenance cost comparison, a tabulation of centrifuge construction costs, and an equivalent uniform annual cost comparison were made. The information collected and tabulated is presented in the following. *Note:* Although the Alpha Plant does not dewater biosolids for use in the composting process (it incinerates biosolids and recycles its biosolids ash), this study has relevance for plants that supply biosolids for composting.

CENTRIFUGE CONVERSION STUDY

A. Operation and Maintenance Cost Comparison
   1. Electrical–The following assumptions were made:
      a. One centrifuge and one 40 HP variable-speed positive displacement pump (180 GPM) would be needed in Phase I.
      b. The biosolids transfer pumps would be replaced with the above-mentioned variable-speed pumps, thereby reducing the pumping requirements.
      c. The following equipment would be eliminated:
         - Biosolids conditioning tank mixer
         - Lime feed equipment
         - Live bottom bin drives
         - Filter press drives
         - Constant-speed biosolids feed pumps
      d. The power requirements of the ferric chloride feed system would be equivalent to the power requirements of the new polymer feed system (and therefore cancel out).

      Based on these assumptions, the horsepower comparison between the centrifuge installation and the pressure filter press installation at the current flow rate of 50 GPM are listed below:

- Centrifuge dewatering 2.10 HP/GPM
- Plate and frame dewatering 0.52 HP/GPM

At the current loading, this would increase Alpha Plant's electrical cost by $22,700.00.

2. Major Repairs and Replacements–The following assumptions were made:
   a. The average of the major repair and replacement costs at other plants using centrifuges is indicative of what the cost would be at Alpha Plant.
   b. The comparison of the last six years' costs would minimize the unusually large expenses that might occur on the odd budget year.
   c. The effect of changing budget categories or procedures would not have a major impact on the comparison.
   d. Knowing the age or model of centrifuge used at the different facilities would benefit the comparison, because it would indicate equipment upgrade costs as well as the cost of repairs to older equipment. The total major repairs and replacement costs for four categories of equipment installed at five similar wastewater treatment plants (four with centrifuges and Alpha Plant pressure filter presses) are listed in Table 2.3. These totals are for the six-year timeframe: 1986–1992.

   The average of the four plants with centrifuges is $192,500. Alpha's costs:

$$\frac{\$791{,}500}{192{,}500} \times 100 = 411\% \text{ higher on average}$$

The average improvement fund cost savings would be

$$\frac{\$791{,}500 - 192{,}500}{6 \text{ years}}$$
$$= \$99{,}800/\text{year}$$

3. Chemical Costs–The following assumptions were made:
   a. Future changes in polymer and in lime and ferric chloride pricing be similar (on a % basis).
   b. The worst-case projected polymer dosage is accurate.

   The projected chemical cost comparison is listed as follows:

TABLE 2.3. Major Repairs and Replacement Costs for Four Categories of Equipment at Five Plants.

	Plant #1	Plant #2	Plant #3	Plant #4	Alpha Plant
Chemical feed	56,000	39,500	35,400	61,200	209,700
Feed pumps	2,000	27,800	20,400	15,900	185,100
Dewatering equip.	181,400	105,900	75,400	76,700	200,200
Biosolids conveyance	7,400	40,700	20,300	3,300	196,500
Total	246,800	213,900	151,500	157,100	791,500

- $29.91/ton for ferric chloride and lime
- $27.00/ton for polymer

This cost difference equates a projected savings of $2.91/ton. If 6,820 tons are processed each year as projected, the cost saving would be $20,000/year. In addition to the above chemicals, the need for muriatic acid would be eliminated; this equates to $30,600 or $4.50/ton.

4. Hauling/Landfill Costs—The following assumptions were made:
   a. The lime and ferric chloride added to the biosolids passes through the process into the ash system with no reduction in weight.
   b. The polymer added to the biosolids is volatile and adds no weight to the ash.
   c. The headworks' residuals comprise a small enough portion to be ignored (approximately 5% of total).

   The current practice is to add 30% or 600 pounds of chemicals per ton of biosolids. Elimination of this nonvolatile content would directly reduce hauling and landfill costs by 30%. Therefore, current projected savings would be:

$$0.3 \times \$234,400 = \$70,300/\text{year}$$

or

$$\$70,300/6,820 \text{ ton} = \$10.31/\text{ton}$$

5. Personnel—The following assumptions were made:
   a. The incinerator operator would not be able to perform the additional workload of operating the centrifuge, gravity belt thickener building, odor control station and gravity thickener.
   b. The dewatering plant operator assistant's duties would be decreased significantly so that one solids handling maintenance operator assistant could perform housekeeping during dayshift.
   c. The sewage treatment plant operator assistant's duties could be transferred in the following ways:
      - The incinerator assistant would perform housekeeping duties in the headworks.
      - The incinerator assistant would handle residuals movement from the headworks.
      - The dewatering operator would perform all the sampling, testing, and data recording duties of the sewage treatment assistant.
      - The sewage treatment operator assistant's housekeeping duties would be handled by an assistant sewage treatment maintenance operator assistant assigned to dayshift.
   d. The dewatering plant operator would be responsible for the operation of the:
      - primary biosolids
      - intermediate biosolids
      - RAS and WAS

- gravity belt thickeners
- gravity thickener
- odor control station "D"

e. The reduction in preventive maintenance is equally divided between maintenance operator and maintenance operator assistant manhours.

Note: The preceding information relates to Alpha Plant's present requirement of having to utilize a separate group of shift workers to operate the pressure filter press system for dewatering and batch processing, which is very labor intensive.

The results of this study indicate that four sewage treatment plant operator assistants, four solids handling plant operator assistants, and one solids handling maintenance operator positions could be eliminated.

Among the reduced work force, the duties would be divided as suggested in the assumption such that the sewage treatment operator assistant's duties would be divided between the incinerator operator assistant, the dewatering operator, and the new position of sewage treatment maintenance operator assistant. The solids handling operator assistant's duties would be greatly reduced with the new dewatering process, such that they could be assumed by the reassigned solids handling maintenance operator assistant.

One sewage treatment maintenance operator assistant would be added (as mentioned earlier, this position would essentially be a housekeeping duty). Table 2.4 shows current and proposed staffing of the affected positions.

Cost reductions per year (based on FY 92–93 personnel costs) would be:

Salary

Plant Operator Assistants	$21,806 × 7 =	$152,642[1]
Maintenance Operators	$32,110 × 1 =	$32,110[1]
	SUBTOTAL	$184,752[1]

TABLE 2.4. Current and Proposed Staffing of Affected Positions.

	Current	Proposed	Change
Sewage Treatment Plant Operator Assistant	4	0	−4
Sewage Treatment Maintenance Operator Asst.	4	5	+1
Solids Handling Plant Operator Asst.	8	4	−4
Solids Handling Maintenance Operator	4	3	−1

[1]Salary figures are not reflective of any one particular entity, instead they are typical of the industry when averaged.

Salary & Benefits

(184,752) + (184,752 × 0.3813)[1]
TOTAL = $255,198[1]

B. Centrifuge Construction Costs
Listed below is the equipment required and the estimated construction costs:

Centrifuges	$2,025,000
Removal of existing transfer pumps and replacement with owner-furnished variable-speed pumps	18,000
Piping to centrifuges	33,700
Base supports for centrifuges	45,000
Furnishing and installing stainless steel belt conveyors (with odor control)	337,500
Electrical power and control for conveyors	6,000
Polymer feed system	37,500
Access to conveyors	45,000
Incinerator feed screw hopper modifications	6,700
Incinerator cake feed conversion	15,000
TOTAL	$2,569,400

C. Equivalent Uniform Annual Cost Comparison (EUAC)
Below is a comparison of the annual cost–difference between a centrifuge facility and a plate and frame filter press facility. The centrifuge facility is projected to reduce O and M costs by $473,300; therefore, the O and M cost of the pressure filter press was assumed to be $473,300 (for simplicity).

Centrifuge EUAC 10 YEARS
= 2,569,400 (A/P, 6%, 10)
= 353,180
Pressure filter press = 473,300
Centrifuge EUAC 8 YEARS
= 2,569,400 (A/P, 6%, 8)
= 418,607
Pressure filter press = 453,200

On completion of Alpha Plant's centrifuge conversion study, results were discussed with key plant personnel during a MTQ planning session. The plant MTQ team recommended forwarding the study and results to higher technical authority for critical analysis.

The results of the critical review were as follows,

According to the district engineer who reviewed the study, the pilot test conducted at the Alpha Plant to determine the performance of centrifuging biosolids, and thus the feasibility of changing over from pressure filter press dewatering to centrifugation, produced a 27% total solids (TS) cake while using 15 pounds per ton of polymer.

The main cost savings were as follows:

- decrease in improvement fund cost by approximately $100,000 per year
- decrease in chemical cost by approximately $50,000 per year
- decrease in transportation by approximately $70,000 per year based on current flows
- decrease in personnel cost by approximately $225,000 per year

A slight increase in electrical cost of $23,000 per year is expected due to the higher power usage of the centrifuges. In addition to the cost savings, an increase in incinerator capacity would be realized due to the elimination of the current chemical conditioning that results in 30–35% inert solids being fed to the incinerator.

Based on the information presented in the Alpha Plant manager's report and the analysis of the report by the district engineer, a recommendation to approve the changeover from pressure filter press dewatering process to centrifugation was passed on to higher authority.

In this case, "higher authority" was in the person of the district director of utilities. The director of utilities reviewed the study conducted by the Alpha Plant manager and the critical assessment made by the district engineer and made the following assessment and recommendation to the country board of supervisors

---

I have reviewed these data and agree with the projected operating cost savings. I have also reviewed the projected cost and, based on data received from a consulting engineering firm, believe the data should be revised upward to reflect current market cost for centrifuges. This revision results in a construction cost of $2,804,000. If engineering cost at 12% is added, the total project cost becomes $3,140,000.

Based upon equivalent annual cost, this project becomes a cost-effective alternative if the centrifuge life is ten years (note: experience with existing centrifuges has demonstrated a service life of over fifteen years). The project would have a simple payback of seven years. In addition, based on other pressure filter operations, it is anticipated that three of the existing pressure filters will require major rework within the next five years at a cost of approximately $500,000 per filter. Accordingly, I have requested that the replacement of the pressure filters at Alpha Wastewater Treatment Plant be included in the capital improvement budget.

Along with the above written recommendation, the following pressure filter replacement evaluation cost projections were included in the director's report:

Centrifuges 3 @ $750,000 each	$2,250,000
Replacement of transfer pumps	18,000
Piping to centrifuges	33,700

Base supports for centrifuges	45,000
New conveyors (including installation)	347,500
Electrical power and controls for conveyor	6,000
Polymer feed system	37,000
Conveyor access	45,000
Incinerator feed modification	21,700
Subtotal[2]	$2,803,900
Engineering and contract admin @12%	$336,468
TOTAL	$3,140,368
Equivalent annual cost based on 10 years and 6%	$426,675
Simple payback = 7 years	

The county board of supervisors reviewed, discussed, and approved the centrifuge installation project. To date, centrifuges have been installed at the Alpha Wastewater Treatment Plant and the two WWTPs that supply biosolids to a composting facility. Early results indicate operation at levels better than projected, and cost savings appear to be in line with expectations.

DISPOSAL

With or without biosolids thickening, stabilization, conditioning, and/or dewatering, the wastewater treatment facility involved will have a plan or routine to follow in disposing of treated or untreated biosolids.

Biosolids produced during the wastewater treatment process may contain concentrated levels of contaminants originally contained in the wastewater. Because of the EPA's 503 regulations and concern for public well-being, a great deal of concern must be directed toward ensuring the proper disposal of these biosolids. The point is, in ultimate disposal, the goal must not be to merely shift the original pollutants in the wastestream to a final disposal site where they may become free to contaminate the environment.

This text takes the view that there is a more reasonable approach in the ultimate disposal of wastewater biosolids: Biosolids should be looked at as a resource that can be reused or recycled.

As stated previously, all biosolids produced at a wastewater treatment plant must be disposed of in some way. The treatment processes described to this point may reduce biosolids volume or change its characteristics, but still leave a residue which in most cases must be removed from the plant site.

[2]Itemized cost includes contingencies.

Like the liquid effluent from the treatment plant, there are several other methods for the disposal of biosolids: (1) disposal in water (no longer permitted), (2) disposal on land, (3) for use as a fertilizer or soil conditioner, (4) for land reclamation projects, and (5) for composting. This applies whether or not the biosolids is treated to facilitate or permit the selected method of disposal.

The purpose of this text is to address ultimate disposal of biosolids through composting and subsequent beneficial use in land application. Before beginning a discussion of the main topic, biosolids-derived composting, the following disposal methods will be discussed briefly.

Disposal in Water

Until recently, the major alternatives for sewage biosolids disposal were water and land disposal. This is no longer the case. Water or ocean disposal was an economical but seldom-used disposal method because of its contingency on the availability of bodies of water adequate to permit it. At some seacoast cities, it was common practice for biosolids, either raw or digested, to be piped off shore or pumped to barges and carried to sea to be dumped in deep water far enough off shore to provide huge dilution factors and prevent any ill effects along the shoreline.

As time went on and quantities of sewage biosolids dumped into oceans increased, increased pollutional loads, well above safe standards, affecting beaches along the upper East Coast region of the United States became apparent and began to be a cause of concern. The "out of sight out of mind" philospophy of dumping biosolids into the ocean did not work because the ocean-dumped biosolids started to creep back toward the shoreline. For this reason the dumping of sewage biosolids was prohibited by Congress in 1992.

Disposal on Land

Disposal of sewage biosolids on land gained attention when ocean dumping was prohibited and when rigid air quality regulations began having direct impact on incineration of biosolids. This is not to say that regulations do not apply to land-applied biosolids; they do. For example, new federal regulations under CFR 40 part 503 impact the disposition of biosolids using land disposal by the following methods:

(1) Landfilling
(2) Application as fertilizer or soil conditioner
(3) Land reclamation

Landfilling

According to Corbitt (1990), the most widely accepted means of disposal of wastewater biosolids is landfills. This is starting to change, however, as the result of two major influences: (1) Large densely populated areas are running out of space in landfills for any type of waste, and (2) the new EPA 503 regulations have had direct impact on the way in which wastewater biosolids can be disposed of.

When used for fill, wastewater biosolids is confined almost entirely to digested biosolids, which can be exposed to the atmosphere without creating serious or widespread odor nuisances. Biosolids must be well digested without any applicable amount of raw or undigested biosolids mixed with it.

For Fertilizer or Soil Conditioner

In some forms of agriculture, application of human sewage as a fertilizer is nothing new. For example, human and livestock excrement has been used in Monsoon Asia in tropical "garden culture" for years. To a degree, sewage biosolids is the modern counterpart of "night soil." The key phrase, of course, is *to a degree.*

As a case in point, consider that sewage biosolids is an organic source of nutrients. Biosolids containing as much nitrogen and phosphorus as farmyard manure is agriculturally valuable. However, the nutrient content of sewage biosolids varies greatly and is always lower in potassium than is farmyard manure. Making a nutrient value comparison between "night soil" and farmyard manure and sewage biosolids requires another consideration. That is, urban/industrial waste often contains metallic elements that are nonessential and in quantities that can be toxic to plants and animals. Moreover, when these sewage biosolids contain heavy metals and other toxic agents and are applied too frequently and for too long, increases in the levels of such heavy metals as copper, cobalt, boron, lead, mercury, and others may result. Complicating this potentially toxic soil condition is that once these contaminants enter the soil, they cannot readily be removed (Tivy, 1990).

Notwithstanding the potential consequences of using biosolids contaminated with heavy metals (to be discussed later), sewage biosolids used for fertilizing and soil conditioning is a beneficial end use because nutrients are recycled and disposal costs are reduced. The U.S. Environmental Protection Agency (EPA) distinguishes between application of biosolids to agricultural and nonagricultural land (McGhee, 1991). Under the Clean Water Act of 1972 as amended (1978), the EPA promulgated 40 CFR Part 503 Final Rules for Use and Disposal of Sewage

Biosolids. The 503 regulation sets forth a comprehensive program for reducing the potential environmental risks and maximizing the beneficial use of biosolids.

The EPA assessed the potential for pollutants in sewage biosolids to affect public health and the environment through a number of different routes of exposure. Specifically, the agency evaluated the risks posed by pollutants that might be present in biosolids applied to land and considered human exposure through inhalation, direct ingestion of soil fertilized with sewage biosolids, and consumption of crops grown in the soil with sewage biosolids. The agency also assessed the potential risk to human health through contamination of drinking water sources or surface water when biosolids is disposed on land. The final rules were publsihed in the *Federal Register,* February, 1993. For any operator or manager working in the wastewater industry it is important to become familiar with and abide by the 503 regulations.

Application of sewage biosolids to land is a popular disposal method because it is relatively simple to accomplish at a relatively reasonable cost. Biosolids land application is not only a viable disposal alternative, it can also make available many elements essential to plant life, such as nitrogen, phosphorous, and potassium. Additionally, biosolids also provides traces of other nutrients that are considered more or less indispensable for plant growth, such as calcium, copper, iron, magnesium, manganese, sulfur, and zinc. However, these other elements are sometimes found in concentrations, perhaps from industrial wastes, that may be harmful. Another benefit gained through biosolid land application is obtained from its humuslike quality, which, besides furnishing plant food, benefits the soil by increasing its water-holding capacity and improving tillage, thus making possible the working of heavy soils into satisfactory seed beds. Finally, application of biosolids to land helps to reduce soil erosion.

Soils vary in their requirements for fertilizer, but it appears that the elements essential for plant growth may be divided into two groups: those that come from the air and water freely and those that are found in the soil or have to be added at certain intervals. The first group includes hydrogen, oxygen, and carbon. The second group, collectively called the *macronutrients,* consists of nitrogen, phosphorous, potassium, and several miscellaneous elements usually found in sufficient quantities in most soils, such as calcium, magnesium, sulfur, iron, manganese, and others.

The major fertilizing elements are nitrogen, phosphorous, and potassium (these are all in chemically combined forms, and are not present as free elements), and the amount of each required depends on the soil, climate conditions, and crop. Normally, fertilizers used for crops contain nitrogen, phosphorous, potassium percentages in varying amounts

(Vesilind, 1980). For example, 8-8-8 and 5-10-5 blends are commonly used (Oberst, 1996).

The amount of nitrogen, phosphorus, and potassium that is available from sewage biosolids is much lower. This is the case because during the various treatment processes, biosolids loses some of its nutrient value. Still, typical biosolids from municipal wastewater treatment facilities contains a nitrogen-phosphorous-potassium percentage relationship of about 3-2-0 (Henry & Harrision, 1992). A typical example of a final nitrogen-phosphorous-potassium percentage relationship can be seen in a common biosolid-derived composit soil amendment product that has a 2-2-0 percent content (see Figure 2.7). Digested biosolids is somewhat comparable to farm manure in its content of fertilizer constituents, relative availability, and the physical nature of the material.

The value of fertilization can be seen in the benefits obtained from nitrogen. Nitrogen is required by all plants, particularly where leaf development is required. Thus, it is of great value in fertilizing grass, radishes, lettuce, spinach and celery. Moreover, nitrogen stimulates growth of leaf and stem.

Some tests have demonstrated that activated biosolids used as an organic carrier for added inorganic forms of nitrogen has given better results from crops with a short growing season than activated biosolids alone. The in-

Figure 2.7 A 40 lb bag wrapper for biosolids-derived compost with a nitrogen-phosphorous-potassium composition of 2-2-0.

organic nitrogen is quickly avialable while that from the organic portion is available more slowly and lasts over a period of time. Laws (1993) estimates that the fertilizer nitrogen requirements of about 0.35 million hectares (2.5 acre = 1 hectare) of cropland could be supplied with sewage biosolids. While this application of sewage biosolids as a source of nitrogen for agriculture may sound impressive, keep in mind that all the sewage biosolids produced in the United States could supply no more than a small fraction of the annual fertilizer nitrogen required for crop production in the United States.

Like nitrogen, phosphorous is also essential in many phases of plant growth. For example, phosphorous hastens ripening, encourages root growth, and increases resistance to disease.

Finally, potassium is an important factor in vigorous growth. It develops the woody parts of stems and pulps of fruits, Moreover, potassium increases resistance to disease, but delays ripening and is needed in the formation of chlorophyll.

From the preceding it should be apparent that sewage biosolids is not going to change the United States' reliance on commercial fertilizers for growing crops. At present, the cost of transporting the biosolids for land application in a specific locality may not compare favorably in cost competitive terms with commerical fertilizer. However, this is not to say that land application of biosolids does not have some dollar value. For example, when municipalities bear the cost of transporting the biosolids, then biosolids land application can be more competitive with the cost of purchasing commercial fertilizer and then transporting it for use in the same application. What it comes down to is a site-specific tradeoff: What might be cost effective for one agency, might not be for another.

Another way by which land application of wastewater biosolids can be measured in dollar value is demonstrated in the following example.

Example 2.1: The dollar value of 5 to 10 dry tons per acre of a typical anaerobically digested dewatered biosolids applied to farmland can be seen in Table 2.5. The data in Table 2.5 were compiled by the U.S. EPA (1994) on biosolids from metropolitan Washington, DC, that was used on land in 1993. About 75% of the dewatered biosolids produced was used on agricultural land in Maryland and Virginia. The dewatered biosolids were applied to private farmland by private contractors at no change to the farmers.

The U.S. Environmental Protection Agency also provides the data in Table 2.6 on biosolids chemical composition.

The suitability of a particular land area for biosolids application depends on soil characteristics. The most favorable soils have a high infiltration and percolation rate, possess good drainability and aeration, and have a neutral pH (Corbitt, 1990). When these soil conditions are available,

TABLE 2.5. Dollar Value per Acre of Anaerobically Digested Dewatered Biosolids Applied to Farmland.

Nutrient	lb/Acre Applied	Value/Acre
Nitrogen	150	$30.00
Phosphorous	150	$30.00
Potassium	10	$1.00
Copper	7	$14.00
Zinc	10	$12.50
Sulfur	20	$10.00
Lime	1 ton	$28.00
Spreading		$15.00
Total Value*		$140.50

*Value of organic matter is in addition to this total.

Source: U.S. EPA. *Biosolids Recycling: Beneficial Technology for a Better Environment,* EPA 832-R-94-009, June 1994.

TABLE 2.6. Biosolids Chemical Composition.

	Types of Biosolids		
Constituent	Raw	Digested	Activated
Volatile material	60–80	30–60	60–80
Total dry solids	2.0–8.0	6.0–12.0	1.0
Greases and fats	7–35	3–17	5–12
Protein	20–30	15–20	30–40
Ammonium nitrate	1–3.5	1–4	4–7
Phosphoric acid	1–1.5	0.5–3.7	3–4
Potash	0–1	0–4	0.5–0.7
Cellulose	9–13	10.0	7.8
Silica (SiO)	15.0–20.0	10.0–20.0	8.0
pH	5.8	6.5–7.5	7.0
Iron	2.5	5.0	7.0

dried or dewatered sewage biosolids makes an excellent soil conditioner and a good, though incomplete fertilizer, unless fortified with nitrogen, phosphorous, and potassium. Heat-dried, raw activated biosolids is the best biosolids product, both chemically and hygienically, although some odor may result from its use. By comparision, heat-dried, digested biosolids contains much less nitrogen and is more valuable for its soil-conditioning and building qualities than for its fertilizer content. For some crops it is deleterious. It is practically odorless when well digested.

Biosolids cake produced in vacuum filtration cannot be readily spread on land as a fertilizer or soil conditioner because of its pasty nature. It must be further air dried. At some wastewater treatment facilities the biosolids cake is stockpiled and "wintered over" on the plant site. Wintering over allows for the alternate freezing and thawing and air drying that results in a material that breaks up more readily.

The application of raw biosolids to fields has sometimes been detrimental because the grease and fat content was difficult for the soil to absorb and caused it to become impervious. However, this problem has been greatly reduced by digestion of biosolids whereby the grease and fat is decreased and so finely divided that the porosity of the soil is not adversely affected. However, the continued use of digested biosolids tends to lower the pH value of soil. Consequently, for proper pH adjustment, it is recommended that either ground limestone or lime be applied occasionally.

To this point, we have listed the importance of soil conditions and the nutrient value available in biosolids that is to be applied to land as two significant factors that must be taken into consideration. Another significant factor must be considered: heavy metals. The principal concern is to keep the amount of metals in biosolids at nontoxic levels so that the end product, the crops, contain nontoxic levels. Moran et al. (1986) point out that "although metals from biosolids do not accumulate in seeds, they do accumulate in other plant tissues. Usually, cadmium levels are the biggest worry, because cadmium is a carcinogen" (p. 262).

As stated previously, heavy metal contamination brought about by the use of biosolids in agricultural applications is a potential problem. The EPA addresses this topic in detail in its 503 regulation. Table 2.7 shows the wide variation that can occur in the concentration of heavy metals in wastewater biosolids. It is interesting to compare the range, median, 503 pollutant limits that have no metal-related loading rate restrictions for sewage biosolids and the average biosolids pollutant concentrations for 208 POTWs that participated in the 1987 National Sewage Sludge Survey (NSSS). The additional 503 requirements relating to land application of sewage biosolids (pathogen-reduction requirements and vector attraction reduction) will be discussed later.

TABLE 2.7. Variation in Concentration of Heavy Metals in Wastewater Biosolids.

Metal	Dry Biosolids, mg/kg			
	Range	Median	503 Reg. Limits*	NSSS Mean**
Arsenic	1.1–230	10	41	9.9
Cadmium	1–3410	10	39	6.9
Chromium	10–99,000	500	1200	119.0
Copper	84–17,000	800	1500	741.0
Lead	13–26,000	500	300	134.0
Mercury	0.6–56	6	17	5.2
Molybdenum	0.1–214	4	18	9.2
Nickel	2–5300	80	420	42.0
Selenium	1.7–17.2	5	36	5.2
Zinc	101–49,000	1700	2800	1202.0

*The concentration of pollutants in biosolids must be below the limits listed. Reference: Table 3 503.13 (b) (3).

**Results obtained from a comparison with the mean concentrations from the 208 POTWs that participated in the National Sewage Sludge Survey (NSSS) conducted in 1987. EPA estimates that approximately 70% of biosolids will meet these limits.

Source: U.S. Environmental Protection Agency (EPA) 40 CFR Part 503 Regulations, February 19, 1993. U.S. Environmental Protection Agency: Environmental Regulations and Technology, *Use and Disposal of Municipal Wastewater Sludge,* EPA 625/10-84-003, September 1984.

For Land Reclamation

To date, the most successful land application of biosolids has been the restoration of strip-mined land (Laws, 1993). Attempting to establish vegetation of such land has proven extremely difficult. This is obvious, considering that, typically, strip-mined land lacks nutrients and organic matter and has low pH, low water retention, and high levels of toxic metals (Sopper & Kett, 1981). However, dramatic restoration of permanent vegetative cover on several plots of strip-mined land following the application of sewage biosolids has been accomplished and is well documented (Sopper, 1993).

In 1993 it was estimated that 15–20% of municipal sewage biosolids was applied to land (Laws, 1993). Applying biosolids to restore lands that have been devastated through strip mining and other activities is not only beneficial but also a very practical means of disposal.

COMPOSTING—AN INTRODUCTION

Since ancient times, those involved with farming have known the value of composting, whereby organic wastes were converted into valuable soil amendments. These amendments were used to prevent erosion, provide nutrients, and replenish depleted organic matter, which was lost through intensive farming (Corbitt, 1990).

In more recent years, the late British scientist Sir Albert Howard conducted studies and experiments in India that established the basic principles of composting. Sir Albert understood that if organic wastes are partially decomposed by bacteria, worms, and other living organsims, a valuable fertilizer and soil conditioner is produced. Sir Albert was ahead of his time when he predicted the need for applying his concepts to large-scale composting of municipal wastes (Cheremisinoff & Young, 1981).

With the environmental movement of the 1970s, people became more aware of the environment. Moreover, concern grew about the possible adverse affects of anthropogenic (man-made) pollutants. As stated earlier, it was common practice to ocean dump huge quantities of sewage biosolids. Along with ocean dumping, biosolids were deposited here and there on landscapes throughout the United States. Landfilling biosolids was also a popular ultimate disposal option.

However, the problem was that all of these disposal methods were accomplished with little thought to the possible harmful affects to the ecosystems. Moreover, eventually it was recognized that when sewage biosolids were ocean dumped several miles off shore, the same biosolids would find their way back on shore or destroyed or adversely affected the ocean ecosystem. Thus, other means of biosolids disposal had to be found.

The focus for biosolids disposal shifted: ocean dumping was out and landfilling was in. But another problem arose: landfills started to fill up, making landfill space limited. Along with limited landfill space, another problem was soon recognized; that is, regulatory agencies, environmentalists, the scientific community, and the public became concerned about landfilling biosolids that might contain pollutants.

It is not difficult to understand how concerns about pollutants in biosolids (with the possible consequences of environmental contamination) provided the motivating force behind the U.S. EPA's development of regulations applicable to disposal through incineration, land application, and ultimate land disposal of biosolids; thus, the EPA's 503 rules were devised and promulgated.

With the concern for clean air driving the movement toward decreased construction of incinerators for the disposal of biosolids and the increase in regulatory requirements (thus increasing costs) for incinerators already on line and incinerating biosolids for ultimate disposal, a search for alternatives resulted.

One alternative for biosolids disposal already mentioned is landfilling. As stated, producing a safe biosolids product for eventual safe deposit in a landfill is not the problem. The problem is that present landfills are full or filling quickly. Compounding the shortage of landfill space is the problem with opening new landfills. The public does not want landfills in their backyards—the "Not in My Backyard" (NIMBY) mindset.

With most alternate biosolids disposal methods either outlawed, unwanted, and/or declared unsafe, Sir Albert Howard's prediction that composting would become a viable option for waste disposal has become a reality. That is, the future Sir Albert was talking about is here—today. It is interesting to note that as early as 1850 sanitarians in the United States recommended land application of sewage. Moreover, by 1873, "the conversion of excremental waste and refuse was engaging the attention of intelligent minds throughout the world" (Tarr, 1981, p. 38).

Today we know that proper treatment methodologies such as heat drying, treatment with alkaline materials, and composting can convert biosolids into useful products that can be considered of "exceptional quality" if pollutant concentrations do not exceed the minimum levels specified in Table 3 of the Part 503 rule (U.S. EPA, 1993). Moreover, these products are safe for unrestricted use by the general public.

An example of these stabilized products is Hampton Roads Sanitation District's composted biosolid product, "NUTRI-GREEN," which has been produced and marketed successfully in Virginia for several years (see Figure 2.7).

COMPOSTING DEFINED

Earlier in this text compost was defined as the end product (innocuous humus) remaining after the composting process is completed. The composting process can be further defined as the aerobic thermophilic biological degradation of organic wastes to compost (the relatively stable innocuous humus) that can be used as a soil conditioner. The composting of sewage biosolids is a controlled process conducted in a designed and constructed site, and is less complex than composting of general municipal waste, since it is relatively uniform in character and in particle size (McGhee, 1991). The wastewater biosolids used in the composting process must be dewatered to 20–40% solids because the composting process operates on solid materials (feedstocks). During the composting process the bulk of the waste material is decreased. Decomposition results in the biological activity of microbes whose nature depends to some extent on the nature of the material being composted (Singleton & Sainsbury, 1994). A quality compost derived from biosolids can contain up to 2% nitrogen, 2% phosphorous, 1% potassium, and many trace elements. The nutrient content of biosolids-derived compost is not its most valuable feature, however. Instead, it is its moisture-retaining and humus-forming properties that are of significance.

Metcalf and Eddy (1991) report that from 1983 to 1988, the number of biosolids-to-compost facilities almost doubled (61 to 115) due to the anticipated shortage of available landfills and in landfill sites. As pointed out by Vesilind (1980), it is important to note that wastewater biosolids, even varieties that are well digested, are not environmentally ready for use on land but must first be stabilized. If not properly stabilized, these biosolids are not only offensive to the senses, but may also contain substantial quantities of chemical toxins and/or pathogenic microbes.

In order to safely apply biosolids to land, the biosolids must be of "exceptional quality," be reasonably safe, and be aesthetically acceptable. These criteria are satisfied when biosolids-derived compost is the option chosen for land application. *Note:* When attempting to learn about a process such as biosolids composting, it is important to know something about sampling and testing procedures that are important parts of the process. In the attempt to gain knowledge about the biosolids composting process two views can be taken: (1) the macro-view or (2) the micro-view. In taking the macro-view, the administrator or plant/process manager is interested in the "big picture"; that is, the process is studied in a cursory fashion. The micro-view, on the other hand, approaches the study with close scrutiny of minute details. In this text both the macro- and the micro-view approaches are used because it is important for decision-makers to make sound decisions based on relevant data.

With the preceding note in mind, the following section (and others to follow) present both the macro- and micro-views related to biosolids-derived composting. When several micro-views (or small pieces of the puzzle) are arranged together, the "big picture" will equal a macro-view of the puzzle.

BIOSOLIDS TESTING PROCEDURES (PRIOR TO COMPOSTING)

As stated previously, various regulations require testing of biosolids. For example, the 503 rule requires that biosolids that are to be land applied be tested for *Salmonella.* In addition to required biosolids testing, such as the *Salmonella* test, other routine tests should also be performed. One such test, the routine testing of biosolids for percent of solids content, and the various ways in which this test can be conducted, is addressed in the following section.

TESTING BIOSOLIDS FOR PERCENT OF SOLIDS CONTENT

During the biosolids dewatering operation, regardless of the type of dewatering process used, it is "good biosolids handling practice" for the biosolids handling operator to take samples of finished cakelike biosolids product (taken from the conveyor after it has been processed and before deposit in storage hopper or holding bin) at least every other hour and to test the sample for percent of solids concentration. Test results should be entered in the daily operating log/record (see Figure 2.8). This procedure is accomplished for a practical purpose. For example, the hourly or every-other hour sample taken from an on-line dewatering process and then tested for its percent solids enables the operator to gauge process performance. This can be very important. For example, in preparing biosolids for ultimate disposal by incineration, the dewatered biosolids cake product can have a significant impact on incinerator operation. This can be seen when one compares the amount of fuel needed to incinerate biosolids cake that is water laden compared to cake that has been thoroughly dewatered. Less fuel is required to incinerate biosolids cake that has a low water content; that is, high solids content means less water to burn off which, in turn, requires less fuel.

Another example of the importance of determining process biosolids cake percent solids content can be seen in the chemical feed process. If, for example, a biosolids dewatered cake sample is taken (with the ultimate end use targeted for composting) and the percent of solids content is 22%

Figure 2.8 Plant operator calculates test results for entry into plant daily operating log.

or greater, then it would be a waste of chemicals and therefore money to add additional chemicals to this product, which has already reached the optimal level of percent of solids.

In the biosolids-derived composting process it is important for the compost facility operator to know the percent solids of the dewatered cake prior to delivery to the composting facility. This information is important because it enables the operator to determine the correct mixing ratio for the bulking-material-to-biosolids mixing process.

In performing routine dewatered biosolids cake sampling and testing for solids concentration, the operator has a number of test protocols to chose from. For example, some plants may use the Denver Instruments moisture analyzer method, while other plants use the laboratory centrifuge or the Ohaus balance methods. In addition, a more exacting test protocol that will render more precise results is provided in the latest edition of *Standard Methods*.

For informational and illustrative purposes, the four test methods mentioned earlier as used in determining total solids or percent of solids content in dewatered biosolids cake will be presented in the following. It is not the intention here to recommend one procedure over another. Instead, we will point out and illustrate methodologies that are in use in wastewater treatment facilities.

It should be pointed out that the first three tests are designed to be performed by the operator in his/her work station (the treatment plant). Typically, grab samples are taken, solids percent content is measured, and the system is adjusted or left to operate as is, depending on the test results. On the other hand, the fourth test, the protocol described in the following, adapted from *Standard Methods,* is designed to be performed in the wastewater treatment plant laboratory by qualified laboratory technicians.

For the solids handling operator who is interested in taking a biosolids cake grab sample to perform a percent solids content test that will render a relatively accurate estimation of the percent solids concentration of dewatered biosolids cake for process optimization, the Denver Instruments moisture analyzer, laboratory centrifuge, and Ohaus balance methods are user-friendly tests that are quick and easy to perform.

Denver Instruments Moisture Analyzer Method

As the name suggests, the Denver Instruments moisture analyzer method uses a Denver Instruments IR-100 moisture analyzer with disposable weighing pans, spatulas, and tweezers [see Figure 2.9(a) and (b)].

After taking a biosolids sample (see Figure 2.10), the operator turns on the IR-100. Since this instrument takes about thirty minutes to warm up, usual practice is to leave it in the ON mode.

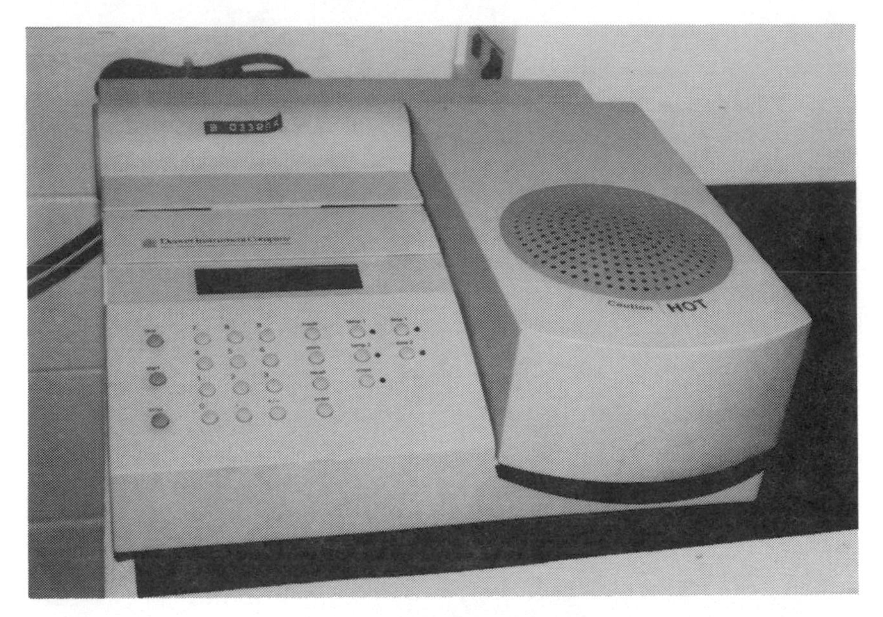

(a)

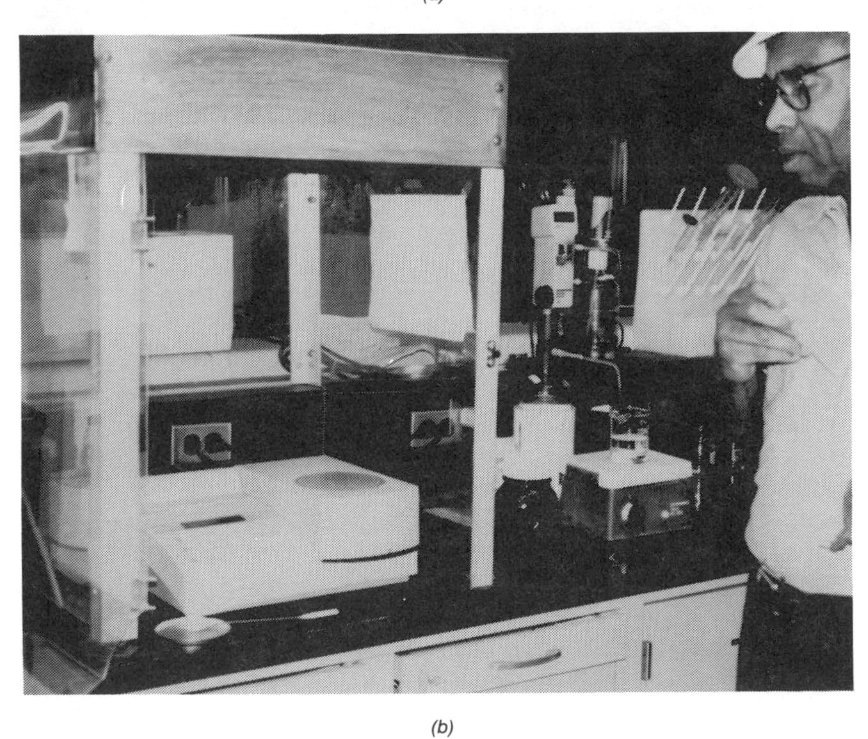

(b)

Figure 2.9 (a) Denver Instruments IR-100 moisture analyzer. (b) Solids handling operator, Denver Instrument (inside vent hood) with associated equipment used for determining percent solids concentration of biosolids sample.

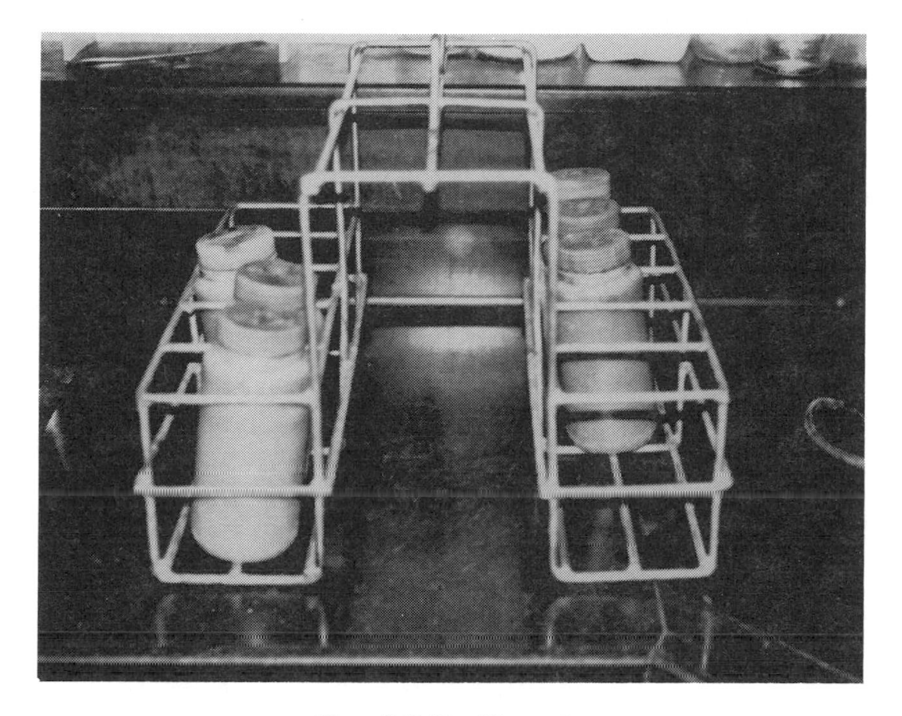

Figure 2.10 Biosolids samples.

After ensuring the IR-100 is warmed up, calibrated, and ready for use, the operator places approximately 1.0 gram of well-mixed sample on the weighing pan. Then, with a spatula, the operator evenly spreads a thin layer of sample on the pan surface (see Figure 2.11).

After placing biosolids sample into weighing pan, the pan is placed in the IR-100 (see Figure 2.12). It may be necessary to evenly spread the thin biosolids layer again. The instrument weighs the sample in a few seconds and displays the sample weight and "PRESS START." Immediately after the weight stabilizes, the operator presses the "start" key and the display shows:

WT = sample weight
WORKING

Several seconds later, after the drying cycle ends, the IR-100 displays the following (*Note:* The display continuously updates data (%TS) as time progresses):

I. WT = Initial weight of sample
F. WT = Final weight of sample

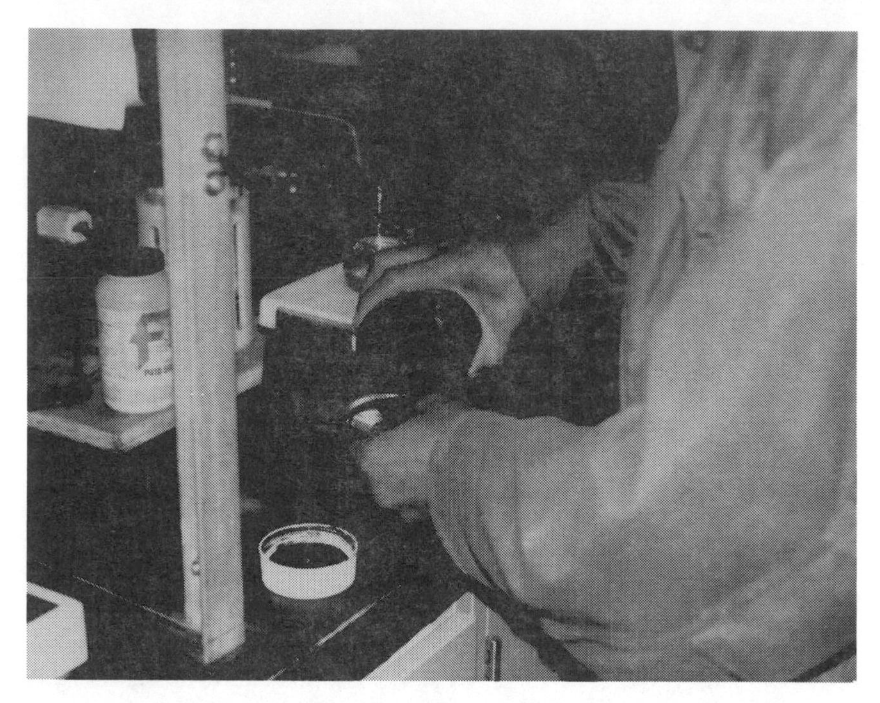

Figure 2.11 Placing biosolids sample into weighing pan.

Figure 2.12 Biosolids sample is placed in IR-100.

Then the operator changes the display to

PERC = Sample % solids value

the following information will appear and stay on the display:

DONE Time of day
PERC = Sample % solids value

The operator then records the sample percent solids value on the daily plant operating record/log (see Figure 2.8). When the sampling procedure is completed (see Figure 2.13) and the results have been entered in the plant record/log, the sample is removed and utensils are cleaned.

Laboratory Centrifuge Method

In the laboratory centrifuge method of determining percent solids concentration, the operator uses a standard laboratory centrifuge (see Figure 2.14) and centrifuge tubes (graduated in either percent or milliliters (ml).

Figure 2.13 Solids left in sample after IR-100 procedure.

Figure 2.14 Standard laboratory centrifuge used for testing percent solids concentration.

The centrifuge procedure the operator follows is listed as follows:

(1) He/she collects a fresh biosolids cake sample.
(2) The operator mixes the sample thoroughly.
(3) He/she then fills a clean empty centrifuge tube to the full mark.
(4) The operator places the centrifuge tube into a tube shield on the centrifuge and records the sample name and corresponding centrifuge position number (see Figure 2.15).
(5) The operator places an identical centrifuge tube filled to the full mark with either an additional sample or water opposite each centrifuge tube and records the sample name along with corresponding centrifuge position number.
(6) The operator then sets the timer control dial on five minutes and the speed control dial to the maximum speed setting.
(7) After waiting for the period to expire and then engaging the centrifuge brake to slow down the centrifuge, the operator waits for the centrifuge to come to a complete stop.
(8) The operator reads and records the centrifuge biosolids volume

expressed as either percent or milliliters (ml), depending on the type of centrifuge tube using the biosolids-liquid interface as the reference point.

(9) The operator then calculates the percent solids concentration:
  a) For a centrifuge tube graduated in percent:

$$\%\ \text{Solids Concentration} = \%\ \text{Centrifuged Biosolids Volume}$$

  b) For a centrifuge graduated in milliliters (ml):

$$\%\ \text{Solids Concentration} = \frac{\text{ml Centrifuged Biosolids Vol} \times 100\%}{\text{ml Initial Centrifuge Biosolids Vol}}$$

Ohaus Balance Method

In this method, an Ohaus moisture determination balance and aluminum pans are used. The operator first turns on the power to the Ohaus Balance and then calibrates it. After calibration, a fresh grab sample is mixed thoroughly. Then 10 grams of the sample is placed onto the pan us-

Figure 2.15 Tube shields inside centrifuge.

ing the gram scale and vernier. The heat lamp is then positioned 1 1/2 inches above the pan. The heat control dial is set for approximately 4.5 and the timer control dial on approximately twenty-five minutes. Later the operator reads the percent moisture loss using the percent scale and vernier. The operator next calculates the percent solids concentration as follows:

$$\%\ \text{Solids Concentration} = 100\% - \%\ \text{Moisture Loss}$$

Note: This test compares the dryness of two or more similar samples when tested in exactly the same manner. The test is not as accurate as a laboratory analysis for determining solids concentration.

Total Solids in Solid Samples Laboratory Method

The following percent total solids laboratory test procedure was adapted from *Standard Methods* (1992) and has widespread use. Unlike the Denver Instruments, centrifuge, and the Ohaus methods, the total solids in solids samples protocol described in the following renders results that are somewhat more accurate. However, this method does not allow for testing with almost instant results on grab samples. This method determines the total solids and its fixed and volatile fractions in solids and semisolid samples such as sediments, various treatment process samples such as centrifuge feed, centrate, and in dewatered cake (the focus of this discussion). The method is subject to negative error due to loss of ammonium carbonate and volatile organic matter during drying. After drying or ignition, residues are very hygroscopic and rapidly absorb moisture from the air.

1. Sample Handling and Preservation
   A. Container type
      Collect samples for total solids and total volatile solids in resistant glass or plastic bottles, provided that the material in suspension does not adhere to the container walls.
   B. Interferences
      Shake or stir the sample thoroughly to ensure that it is well mixed. Disperse any visible floating oil and grease before pouring. Large particles or nonhomogeneous masses are to be excluded from the sample with the approval of a chemist or senior laboratory technician, and noted on the worksheet.
   C. Preservation
      Keep samples at $\leq 4°C$.
   D. Holding time
      The holding time is seven days including the time of collection.

2. Apparatus and Equipment
   A. Apparatus
      - evaporating dishes, 100 ml, porcelain, platinum or high silica glass
      - muffle furnace, 550 ± 50°C
      - drying oven, 103 to 105°C
      - appropriate top-loading balance
      - spatula or spoon
      - crucible tongs
      - timer

   B. Glassware preparation
      Evaporation dishes must be washed with detergent and tap water until clean and rinsed with tap water. Clean dishes must be ignited in a muffle furnace at 550°C ± 50°C for one hour. Dishes must be cooled and placed in a desiccator until used.

3. Procedure
   Note: One set of duplicate analysis must be run for every ten samples or for any batch of less than ten samples analyzed.
   a. Tare the balance, and weigh a clean, ignited evaporating dish. Record the dish number and weight on the data sheet.
   b. Before removing the evaporating dish, tare the balance again.
   c. Remove the dish from the balance.
   d. Shake the sample vigorously and pour the appropriate amount of sample (25 grams for solid or dewatered cake and 50 grams of all other samples) into the dish.
   e. Place the dish back on the balance and add or remove, using a spoon or spatula, sufficient sample to bring the weight to exactly 50.00 ± 0.5 grams for fluid samples or 25.00 ± 0.05 grams for solid samples.
   f. Remove the dish from the balance.
   g. Place the filled dishes in the drying oven overnight at 103–105°C.
   h. When the oven temperature stabilizes (15–30 minutes), after loading the oven with the filled dishes, record the oven temperature on the data sheet.
   i. The following morning, record the temperature on the data sheet before removing the dishes.
   j. Remove the dried samples from the oven and place them in a desiccator.
   k. Allow sample to cool to room temperature.
   l. For each sample, tare the balance, weigh, and record the weight.
   m. Place the dried samples requiring volatile solids determination in the muffle furnance for one hour at 550 ± 50°C.
   n. Remove the dishes and let them cool in the desiccator. Tare the balance, weigh, and record the weights on the data sheet.

   CAUTION: WEAR GLOVES DURING THE PROCEDURE AND PROTECTIVE GLOVES AND GOGGLES WHEN USING THE MUFFLE FURNACE.

4. Results

$$\% \text{ Total Solids} = \frac{A - B}{D} \times 100$$

$$\% \text{ Total Volatile Solids} = \frac{A - C}{A - B} \times 100$$

where:

A = total weight (g) (dried sample and dish)
B = clean dish weight (g)
C = fixed weight (g) (muffled sample and dish)
D = sample weight (g)

SUMMARY AND OTHER IMPORTANT CONSIDERATIONS

This chapter has covered biosolids treatment processes including thickening, stabilization, conditioning, dewatering, and % of solids concentration testing prior to transporting the biosolids cake product to the composting facility for further processing. The next chapter will discuss the capacity and design criteria required for the construction of a biosolids composting facility.

When designing a compost facility it is important to keep in mind that sufficient land area must be available to accommodate:

- heavy equipment
- heavy equipment repair capabilities
- bulking material storage
- mixing area
- composting pad
- curing
- screening
- product storage and distribution pad
- emergency growth areas (contingency, for unexpected surges in biosolids volume)
- materials handling
- administration facility

It is also important to plan for future growth of the composting operation. For example, a study should be conducted to determine how rapidly the surrounding community is likely to grow and, thus, how the community's biosolids generation will increase in the future. The main point to keep in mind is that any newly constructed composting site should include enough growth area to accommodate future growth.

Another important consideration in compost siting is site layout. Factors such as materials flow, traffic patterns, runoff and leachate, and public relations must be addressed. For example, building a compost site that is not easily accessed by delivery and customer vehicles is obviously not smart. Enough room must also be allotted for in-house vehicles. That is, the front-end loaders or other devices used in the composting process must have enough room to move about freely.

Another important area that must be considered is runoff. Will runoff be routed to existing sewer connections or will a separate holding pond have to be built?

Further, public relations cannot be overlooked, especially today. Careful consideration must be given to the public's view. Such elements as site-generated noise, odors, and visual "sore spots" (i.e., the sight of mountains of steamy, smelly biosolids strung along and paralleling a major city highway) should be controlled or avoided. It is also important to ensure that there is enough surrounding space (acreage) to incorporate a "buffer zone" around the composting site.

The planner who diligently pursues a construction plan that includes considerations to make the site community-friendly is on the right track. However, such public relations-oriented planning may not be enough.

If siting becomes a problem, what is the planner to do? One solution may be to site the compost facility next to the POTW. Since locations for many POTWs are remote and removed from domestic centers, siting the compost facility on the POTW property may be prudent.

Another possible site for a biosolids-derived compost facility is the local landfill. Landfills are generally located in isolated areas where odor generation is not always a major consideration. Moreover, landfills can provide amendment material for co-composting. Recent data indicate that the practice of co-composting biosolids with Municipal Solid Waste (MSW) has increased significantly in recent years (Outwater, 1994).

Other factors must also be considered. For example, would local residents object to having biosolids trucked through their community? There have been occasions when trucks transporting biosolids have had accidents. Whenever a truck carrying 24 cubic yards of biosolids is involved in an accident whereby the biosolids is dumped onto a public highway, the potential for a public relations nightmare is high. This is especially the case when such an incident is reported by unenlightened media that tend to incorrectly view biosolids as hazardous waste. Thus, a news report may go out stating that a local POTW has had a serious HazMat spill on a local highway. If this occurs, the public is misinformed and unnecessarily alarmed.

Another question concerning the transportation of biosolids must be addressed. That is, what are the costs of transporting the biosolids? Is the cost reasonable?

At the site itself other questions must be addressed, such as precipitation (rain or snow) and leachate runoff. *Leachate* is the liquid that drains from the compost mix. Precipitation runoff reaches the surrounding pads directly without going through the compost piles. When leachate and runoff are allowed to pool, problems with odor generation and, during freezing conditions, ice formation, which can be dangerous for heavy equipment operation, are likely.

Along with leachate and precipitation runoff, condensate must be considered. Condensate occurs in composting when moisture in the air is pulled through the pile. Therefore, during the design phase, careful consideration must be given to designing a system of collection devices (condensate traps) for installation underneath the piles. In order to prevent pooling of runoff, the compost pad should have a slope of 2–3%. Probably the best way to control runoff is to construct a roof over the compost operation. The ASP model used in this text is less sensitive to moisture problems since only the surface portion, a layer of finished compost, is directly exposed to precipitation.

Probably the most important consideration in the design phase of a biosolids-derived compost facility is odor control. There is no getting around the fact that composting smells. There are a few people around, like the author, who are not offended by this heavy earthy smell; these few seem to be the exception, however. Several factors impact odor generation in composting. These odor factors will be addressed in greater detail later in this text.

Finally, in order to maintain the proper perspective on what is required in the biosolids-derived composting process, the planner and/or the decision-maker should keep the four main goals of composting in mind: 1) "to stabilize the product; 2) to control odor during the process; 3) to get the material dry enough to handle; and 4) to get the temperature high enough to kill pathogens" (Goldstein, 1985, p. 24).

CHAPTER 3

Capacity and Design Criteria

WHEN decision-makers are seriously considering the biosolids-derived compost alternative as a means of ultimate disposal, one of the first things that should be accomplished is to develop a preliminary process design model. This preliminary model should include such factors as the capacity and design criteria of the proposed installation.

CAPACITY AND DESIGN CRITERIA

Material characteristics and design criteria for the illustrative model discussed in this text, the ASP model composting facility, are composites of several inputs. For example, some data presented are derived from an engineering study conducted by Hampton Roads Sanitation District (HRSD) and Black & Veatch Engineering Consultants (1993). Other data represent a combination of information derived from successful biosolids composting operations throughout the U.S. For readers who are contemplating biosolids-derived compost as a beneficial reuse alternative as part of their wastewater operation, the data presented herein are useful because they reflect real-world information that is always required prior to beginning the design phase of a successful project.

An example of the type of design criteria that should be collected is presented in Table 3.1. Although the data in Table 3.1 are intended for illustrative purposes only, they reflect a crucial element that is part of the preliminary process design for any similar facility; that is, data collection. When reviewing this material, keep in mind that a capacity and design criteria study must begin with data related to the design capacity of the wastewater treatment facility or facilities that will provide the biosolids to the future composting facility.

Note: The ASP model composting facility is rated at a capacity of 122.5 dry tons of biosolids per week (dptw). This capacity is equivalent to a 17.5

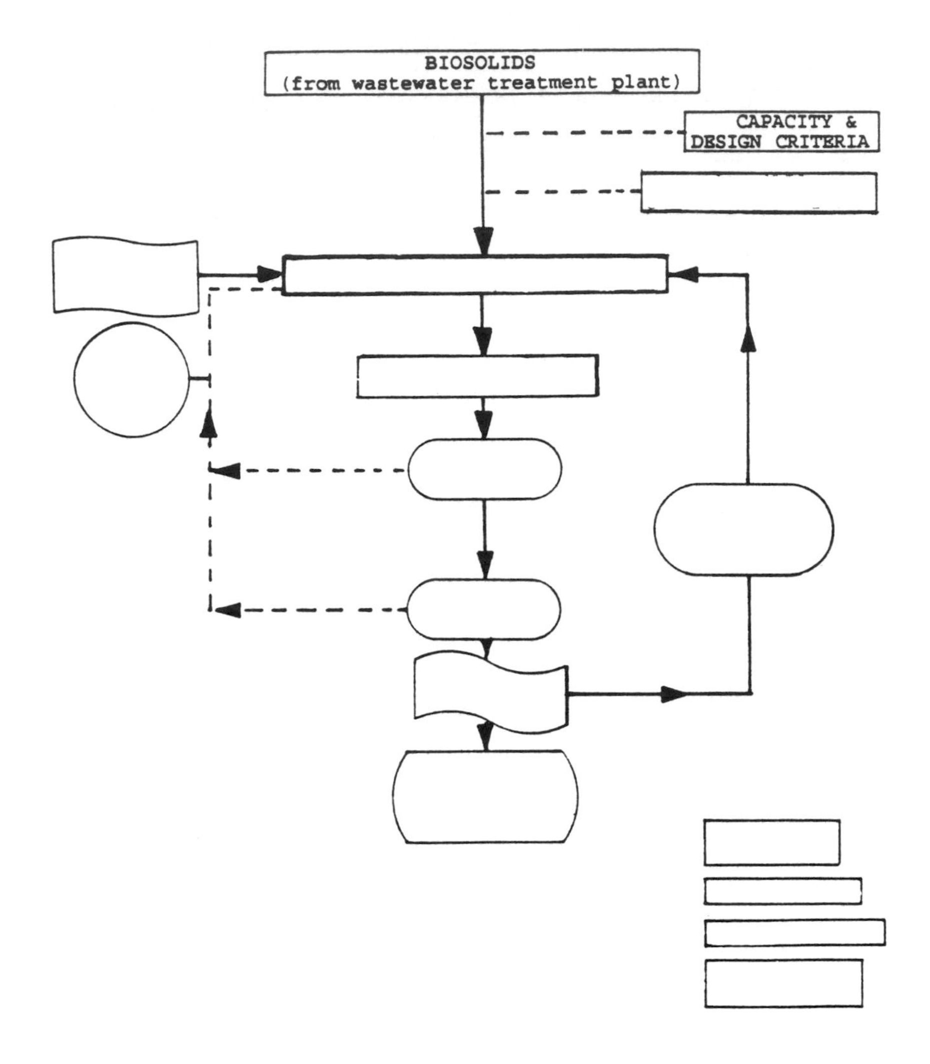

dry tons of biosolids per day (dtpd) on a calendar-day basis. Biosolids, amendment (new and recycled woodchips), compost mix, and final compost characteristics are based on previous experience and field testing (see Table 3.1).

MASS AND ENERGY BALANCES

A series of mass balances must be prepared to define material quantities required for facility design. Material characteristics and process design criteria considered in the mass balances should include the following:

- bulk density of the biosolids, amendments, and initial compost mix
- total solids concentrations of the biosolids, amendment, and mix
- mix ratio of amendment to biosolids
- composting and curing cycle time

These parameters can vary from day to day and can affect daily operations. The mass balances provide a calculation tool for variations in material characteristics so that facilities and equipment can be sized to

TABLE 3.1. Compost Process Design Criteria for ASP Model Composting Facility.

Item	ASP Model Compost Facility
Biosolids	
Quantity, dtpd (calendar day)	17.5
Quantity, dtpw	122.5
Total solids, %	25
Volatile solids, % of TS	63
Biodegradable volatile solids, % of VS	50
Bulk density, lb/cy	1,350
Amendment	
Recycled chips	
Volumetric ratio, amendment/biosolids*	2.5:1
Total solids, %	55
Volatile solids, % of TS	90
Biodegradable volatile solids, % of VS	10
Bulk density, lb/cy	750
New chip	
Total solids, %	60
Volatile solids, % of TS	95
Bulk density, lb/cy	500
Compost Mix Characteristics	
Total solids, % (target)	40
Bulk density, lb/cy	900
Compost Characteristics	
Finished total solids, %	55
Cycle time, days	26
Curing Characteristics	
Initial total solids, %	50
Final total solids, %	55
Cycle time, days	30
Bulk density, lb/cy	900

*Does not include new chips for compost bed and curing pile preparation.
Source: Hampton Roads Sanitation District (1993), used with permission.

ensure that adequate capacity is provided. Various components of the ASP model composting facility are based on different biosolids production conditions. For example, the composting and curing piles are designed based on average month conditions (122.5 dtpw), as is the storage design (122.5 dtpw). However, material handling, such as mixing and screening operations, are sized based on peak week conditions (185.5 dtpw).

Using the capacity and design criteria mentioned earlier, mass balances were prepared for a range of parameter values that might be expected during facility operation:

- biosolids production rates—average and peak week
- biosolids total solids concentration—22 to 30%
- amendment to biosolids mix ratio—2.0:1 to 3.5:1
- biosolids bulk density—1,350 lb/cy (50 lb/cf)

An example of a mass balance iteration used to determine the volume of initial compost mix is presented in Table 3.2.

In order to demonstrate the impact of varying individual parameters on compost mix volume, iterations are provided in Table 3.3 for biosolids processing rates of average, and peak week conditions, biosolids total solids concentrations (22%, 25%, and 30%), and four different woodchip-to-biosolids ratios (3.5:1, 3.0:1, 2.5:1, and 2.0:1). For example, compost mix volume has been determined for average month and peak week conditions for a range of mix ratios and biosolids total solids concentrations. The results of the mass balances for each of these scenarios are summarized in Table 3.3.

It is evident from Table 3.3 that the volume of compost mix and compost pile size are very sensitive to variations in biosolids total solids concentration. By increasing the total solids concentration from 22 to 30%, the volume of compost mix is reduced by 27%. Similary, an increase of total solids from 22 to 25% corresponds to a 12% reduction in the amount of material to be composted. A total solids concentration of 25% or better can be obtained consistently with the high-solids centrifuges that are installed in the wastewater treatment facilities providing biosolids to the ASP model compost facility. Thus, 25% total solids will be used in this text for illustrative purposes relating to the design and operation of the ASP model. With an increase of total solids concentrations above 25%, there is a net reduction in the amount of material to be composted assuming a constant bulk density. For the purposes of this presentation, a biosolids bulk density of 1,350 lb/cubic yard (50 lb/cubic feet) will be used.

For average month conditions (122.5 dtpw), 25% dry biosolids, 1,350 lb/cubic yard bulk density, and a mix ratio of recycled chips to biosolids of 2.5:1, an initial compost mix volume of 2,601 cubic yard/wk was derived. The volumes of biosolids and woodchips are not additive to determine the

TABLE 3.2. **Example Calculations to Determine Mix Volume.**

Start Analysis with 1 Dry Ton Biosolids	Example
1. Pick TSB	25%
2. $W = 1/\text{TSB}$	1/25 = 4.0 wet tons
3. $V = W \times \dfrac{2{,}000 \text{ lb/ton}}{1{,}350 \text{ lb/cy}}$	$4 \text{ tons} \times \dfrac{2{,}000}{1{,}350} = 5.92 \text{ cy}$
4. Select a volumetric mix ratio from 2.5 to 3.5	Say 3.0:1.0
5. $C = \text{mix ratio} \times V$	3.0 × 5.92 cy = 17.76 cy
6. Select a bulk density of woodchips from 550 to 750 lb/cy	Say 750 lb/cy
7. $B = C \times \text{bulk density}$	$17.76 \text{ cy} \times \dfrac{750 \text{ lb/cy}}{2{,}000 \text{ lb/ton}}$ = 6.66 wet tons
8. Select a % solids for woodchips from 50% to 60%	Say 55% solids
9. $A = B \times \text{TSW}$	6.66 wet tons × 0.55 = 3.66 dry tons
10. $X = 1 + A$	1 dt + 3.66 dt = 4.66 dry tons
11. Wet weight of compost (Y) = $W + B$	4.00 wt + 6.66 wt = 10.66 wet tons
12. $\text{TSC} = \dfrac{X}{W + B}$	$\dfrac{4.66 \text{ dry tons}}{(4.00 + 6.66) \text{ wet tons}}$ = 0.437 or 43.7%
13. Select a bulk density of compost	Say 900 lb/cy
14. $M = \dfrac{W + B}{\text{Bulk density}}$	$\dfrac{4.00 + 6.66 \text{ wet tons}}{900 \text{ lb/cy} \times 2{,}000 \text{ lb/ton}}$ = 23.69 cy/dt

Definitions:
TSB—Total Solids Biosolids
TSW—Total Solids Wet
W—Wet Weight
M—Compost Mix Volume/dt Biosolids
V—Volume
TSC—Total Solids Compost

TABLE 3.3. Compost Mix Variations.*

Woodchip: Biosolids Ratio (by volume)	3.5:1	3.0:1	2.5:1	2.0:1
Solids Content of Biosolids 22%				
Avg. month (122.5 dtpw)				
Initial solids mix (%)	44	43	41	39
Initial compost mix vol. (cy/week)	3,643	3,300	2,956	2,613
Peak week (185.5 dtpw)				
Initial solids mix (%)	44	43	41	39
Initial compost mix vol. (cy/week)	5,517	4,997	4,476	3,956
Solids Content of Biosolids 25%				
Avg. month (122.5 dtpw)				
Initial solids mix (%)	45	44	42	41
Initial compost mix vol. (cy/week)	3,206	2,903	2,601	2,299
Peak week (185.5 dtpw)				
Initial solids mix (%)	45	44	42	41
Initial compost mix vol. (cy/week)	4,855	4,397	3,939	3,481
Solids Content of Biosolids 30%				
Avg. month (122.5 dtpw)				
Initial solids mix (%)	47	46	45	43
Initial compost mix vol.	2,672	2,420	2,168	1,916
Peak week (185.5 dtpw)				
Initial solids mix (%)	47	46	45	43
Initial compost mix vol.	4,046	3,664	3,283	2,901

*Assumptions:
Bulk density of biosolids = 1,350 lb/cy.
Bulk density of recycled woodchips = 750 lb/cy.
Bulk density of initial compost mix = 900 lb/cy.
Source: Hampton Roads Sanitation District/Black & Veatch (1993), used with permission.

volume of compost mix. The mix volume is based on an independent bulk density that should remain relatively stable for a given mix ratio. For the ASP model compost facility the determination of mix volume is based on an estimated bulk density of 900 lb/cubic yard.

In order to maintain good porosity for aeration and to prevent leachate drainage; a minimum solids concentration of 40% is typically targeted for the compost mix. The average month scenario defined above results in a total solids concentration of 42% for the compost mix, which is acceptable. Because of the low moisture content in the initial mix, it may be necessary to add moisture to the mix if and when dust control becomes a problem.

In order to define the mix ratio further and determine the minimum amount of amendment needed to provide energy to the ASP model composting operation, an energy balance should be performed. This balance should be developed based on data in *Compost Engineering: Principles and Practices* (Haug, 1980) pertaining to rates of biological activity and heat evolution.

An energy balance for the ASP model using the material characteristics as inputs, the results showed that a mix ratio of amendment (recycled chips) to biosolids of 2:1 was adequate to maintain the energy for the ASP model composting process. It was determined that new chips would not be necessary to supplement the compost mix for energy purposes. New chips are only needed for the compost pile base and as a supplement to the recovered chips to achieve a minimum of 40% total solids in the compost mix. Since the model is based on energy requirements for water evaporation and heat removal, it does not account for the amount of chips necessary to maintain compost pile porosity. Although a smaller mix ratio (corresponding to less volume of compost mix) is feasible, a conservative mix ratio of 2.5:1 was used to establish compost and curing pile volumes and area requirements.

Having used the results of the mass balance model to determine the initial compost mix volume, the mass balance was completed for the rest of the ASP model composting process using the average month design criteria. Key assumptions are noted below:

- Total compost mass = mass of mix + 1′ insulative layer of unscreened compost + 1.5′ new chip pile base for a total pile height of 12.5′.
- Biodegradation occurs through both composting and curing. Biodegradable volatile biosolids and amendment were assumed to be 50% and 5%, respectively, using typical values presented by Haug (1980).
- New woodchip recovery from screens after accounting for biodegradation and screen efficiency is expected to be 70%.

In order to ensure adequate woodchip supply, a storage requirement of three days is provided for recycled chips at the ASP model compost facility. With a 2.5:1 mix ratio, the volume of recycled chips to be stored is three times the daily amount of recycled chips recovered from the screening process (3 × 322.2 cubic yard = 967 cubic yard). For new chips, the ASP model compost facility stores a 13-week supply to compensate for limited supply during the winter months. It should be pointed out that the volume of new chips necessary for the pile bed is dependent on the dimensions of the compost pile. A volume of approximately 14,000 cubic

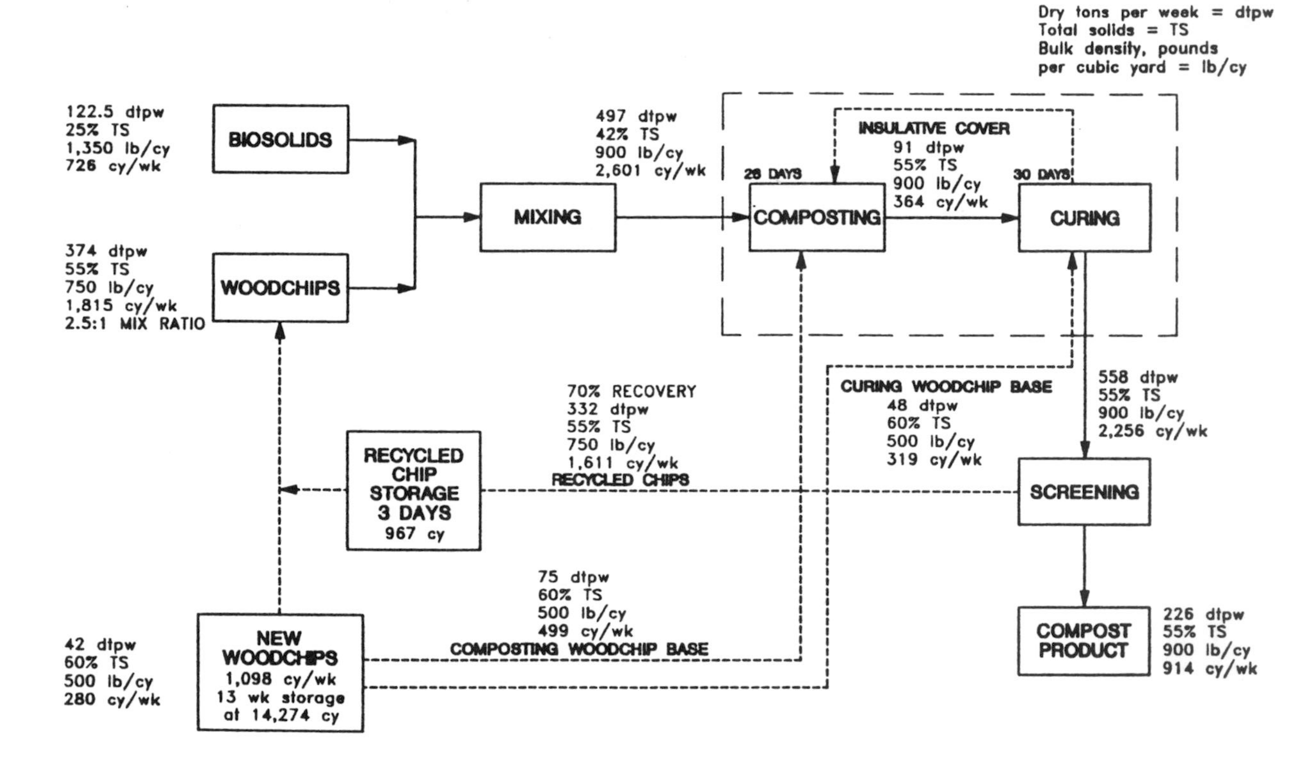

Figure 3.1 Mass balance for ASP model compost facility – shows average design conditions. Source: From a study conducted by Hampton Roads Sanitation District and Black & Veatch (1993); used with permission.

yards of new chips is needed for a 13-week duration based on the pile dimensions for the mass balance presented above.

The mass balance for design conditions at the ASP model compost facility is presented in Figure 3.1. In general, approximately one wet ton of compost is produced per wet ton of compost biosolids.

CHAPTER 4

ASP Model Composting Facility

INTRODUCTION

In this chapter a brief overview is presented of the aerated static pile (ASP) composting facility, the ASP model, and its operation, including background information regarding the site's overall biosolids management plan. Additionally, a brief description of the regulatory requirements governing biosolids composting facility design, construction, and operation is provided.

The ASP model composting facility closely resembles the Peninsula Composting Facility in Newport News, Virginia. This outstanding facility, owned and operated by Hampton Roads Sanitation District (HRSD), is used as the "basic" clone for the ASP model composting facility, because it is a state-of-the-art operation that has won several environmental awards for biosolids composting excellence.

Before beginning our discussion of the ASP model, it would be beneficial to outline and then compare the major types of biosolids-derived composting processes. Along with the aerated static pile (ASP) process, common processes used in composting biosolids include the windrow and in-vessel systems. The windrow composting process entails mixing biosolids with a bulking material/previously dried biosolids and periodically turning the mass. Open windrow systems are usually adequate for digested biosolids but are not suitable for undigested (raw) biosolids. *Note:* A serious problem usually accompanies the composting of raw biosolids: nuisance odors (Corbitt, 1990).

In-vessel composting processes are accomplished inside enclosed vessels or containers. It has several advantages over the other types. For example, the in-vessel system minimizes odors and overall process time by controlling environmental conditions such as temperature, oxygen concentration, and air flow (Metcalf & Eddy, 1991).

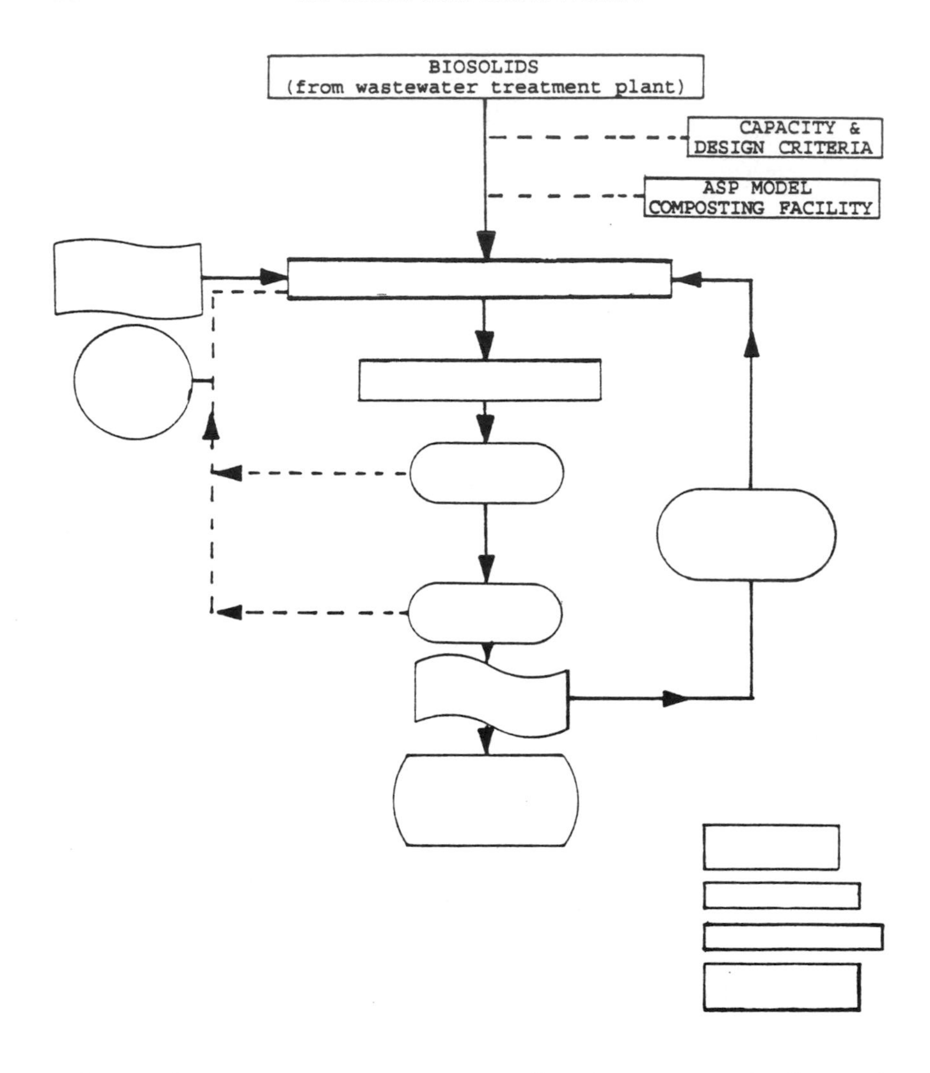

In-vessel composting systems are usually of two types: plugflow and dynamic. The plugflow system is a horizontal system that consists of a totally enclosed bin with a hydraulically operated ram that moves or pushes material flow through the unit. In the dynamic type of in-vessel system, a large-diameter rotating drum works on the biosolids and bulking agents for about 24 to 48 hours (Corbitt, 1990).

In attempting to decide which composting process is best suited for a particular site, it is important to weigh the advantages and disadvantages of each. Figure 4.1 is provided as an aid in making this determination. The matrix makes it relatively easy to determine which process is best suited for a particular or specific operation. This determination can be made

Advantages/Disadvantages	Composting Process Type		
	ASP	Windrow	In-Vessel
High Capital Costs			XXX
Moderate Capital Costs	XXX		
Low Capital Costs		XXX	
High Pathogen Destruction	XXX		
Good Odor Control	XXX		XXX
Good Product Stabilization	XXX	XXX	
High Land Requirement	XXX	XXX	
Operation Weather-Affected	XXX	XXX	
High Labor Requirements	XXX	XXX	
Capacity to handle high Biosolids Volume		XXX	
High Equipment & Maintenance Costs		XXX	XXX
Best Process Control			XXX
Odor Generation Problems	XXX	XXX	
Best Operation Reliability	XXX	XXX	

Source: Data derived from Burnett, C. H. (1992) Small Cities + Warm Climates = Windrow Composting, WEF 65th Annual Conference and Exposition, New Orleans, LA.

Figure 4.1 Comparison of composting types.

based on comparisons of capital cost, degree of pathogen destruction, odor control, product stabilization, and other important considerations such as maintenance requirements and amount of land required.

ASP MODEL COMPOSTING FACILITY

OVERVIEW

The ASP model composting facility was put into operation in 1981. In 1984 an interim projects improvement project was initiated to increase site storage capacity. Further, a compost bagging operation with all necessary equipment and buildings was added in 1986. Demand for compost from this operation has grown at a pace that has resulted in customer demand outpacing the site's ability to supply finished compost product.

To illustrate, consider the trend at the site in the past two years. When the compost product completes the entire composting process and is certified ready for marketing, local buyers literally line the facility access roads with trucks, sometimes stretching for half a mile or more (see Figures 4.2 and 4.3). It is not unusual for this site to sell its entire marketable product in less than one day. Along with its biosolids management and marketing success, the ASP model composting facility is a good friend of the environment. This is evidenced by the site having received the U.S. EPA's National Beneficial Use Award and several regional awards.

The ASP model composting facility's record of successful operation is impressive. Along with employing a beneficial end use for a sometimes

Figure 4.2 Within the composting site a line of customers wait to receive a load of compost.

Figure 4.3 Outside the composting site, a long line of customers forms.

difficult-to-dispose-of waste product, the community served by the local wastewater treatment plants also wins. Local farmers, land reclamation authorities, nurseries, and the public all win. More importantly, when wastewater biosolids-derived compost is produced and utilized correctly, the environment comes out the big winner.

The ASP model composting facility receives anaerobically digested dewatered biosolids from two wastewater treatment plants (WWTPs). (Biosolids may also be received from other local WWTPs, if necessary.) Biosolids are composted at the rate of 17.5 dry tons per day (dtpd). Biosolids are composted using the aerated static pile (extended pile) method to produce a pathogen-free, humuslike material that can be beneficially used as an organic soil amendment. The final compost product is successfully distributed and marketed under the registered trademark name Nutri-Value (not to be confused with HRSD's Nutri-Green). A schematic diagram of the existing composting process is presented in Figure 4.4.

The ASP model composting facility is operated five days per week, Monday through Friday, eight hours per day, and is open for compost product sales and distribution. Biosolids are transported from the WWTPs by Ram-E-Jec trucks and are mixed with woodchips using front-end loaders to obtain a compost mixture with approximately 40% total solids concentration. This mixture is then added to the composting pile and aerated using blowers to provide the air necessary for the biological degradation process.

After twenty-eight days, the mix is removed from the compost pile and

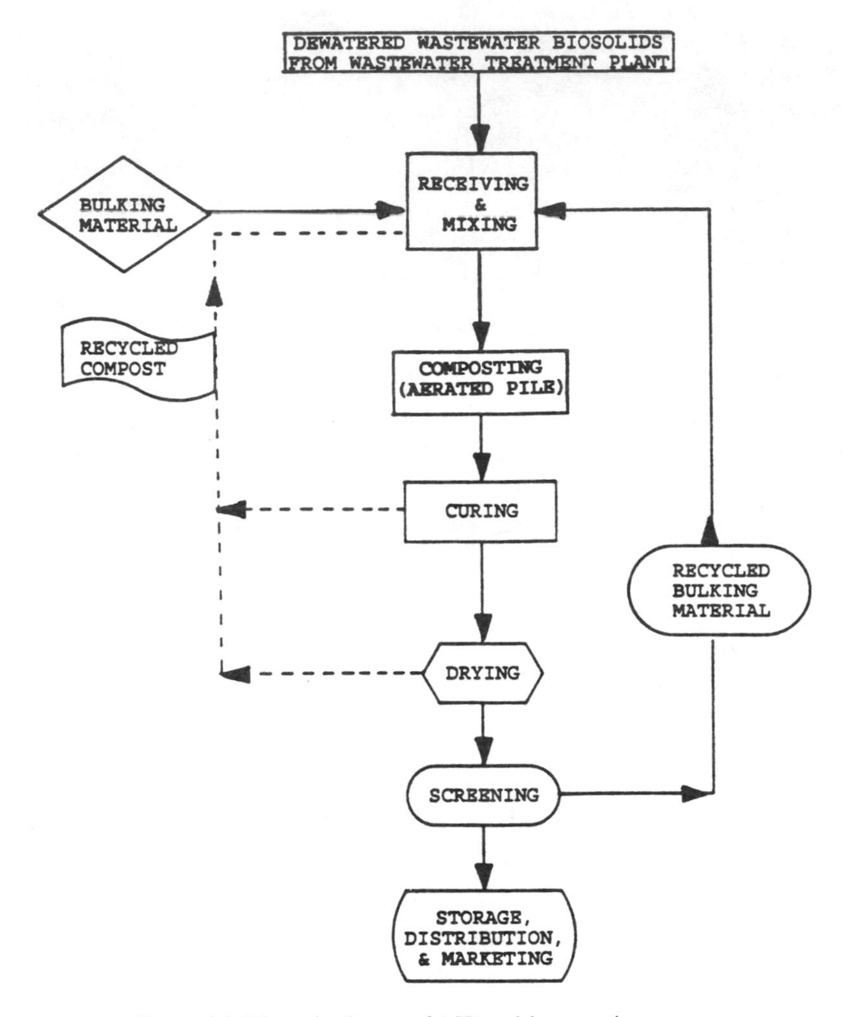

Figure 4.4 Schematic diagram of ASP model composting process.

moved to a separate curing pile. Curing takes place over another thirty days in order to dry the material prior to screening. Screening separates the woodchips from the composted biosolids and produces a marketable product with uniform texture. The woodchips recovered from the screening process are recycled and mixed with the incoming biosolids. The screened final product is stockpiled for bulk sales. A small portion of the product is bagged on site and sold to retailers for resale for home and garden use.

DESCRIPTION OF ASP MODEL COMPOSTING FACILITY

The ASP model composting facility is located on 44 acres of land owned by the local sanitation district. In addition to the administration and maintenance buildings, six pads (bagging and mixing/screening pads have enclosed structures; others are open) are used as operational areas: (1) biosolids receiving and mixing; (2) composting; (3) curing; (4) screening; (5) bagging; and (6) bulk storage for distribution. A layout of the facility is provided in Figure 4.5.

BIOSOLIDS RECEIVING AND MIXING

Biosolids from the wastewater treatment plants is delivered in tractor trailer, Ram-E-Jec trucks with a capacity of approximately 24 cubic yards per load (see Figure 4.6). During a normal work day approximately three truckloads are received from one plant and five truckloads from the other plant. Prior to the trucks arriving, the compost facility chief operator is informed of the total solids concentration of each load so that the appropriate mix ratio of bulking materials to biosolids can be determined. (It should be noted that once each month a biosolid sample is taken at each plant for professional laboratory analysis to determine if *Salmonella* is present. This test is conducted to ensure compliance with the 503 rule.)

When each truckload arrives at the composting site, and after the site chief operator has been informed of the solids concentration of each load, the biosolids is discharged onto a layer of bulking material and mixed with a front-end loader under a canopy area. When weather conditions permit, the mixing operation is conducted outside the canopy area on an open pad. The bucket capacity of the front-end loaders is approximately 10 cubic yards. Additional bulking material is added to achieve the desired mix ratio (to be discussed in detail later).

Prior to the installation of new high-solids centrifuges at the two WWTPs, biosolids had a total solids concentration of about 18%. With the high-solids centrifuges, the biosolids have been dewatered to approximately 25% total solids. Beyond the cost savings anticipated from the changeover from filter-press and belt-press dewatering to high-solids centrifuges pointed out earlier, it is probably safe to say that the dewatered biosolids obtained through centrifuging may produce a cake of 25% or greater total solids on a consistent basis.

In order to achieve a compost mix with at least 40% total solids for the composting process, recycled bulking materials are added at an average 3:1 ratio (bulking materials to biosolids) on a volume basis. The mix ratio is determined based on estimates of biosolids and bulking material quan-

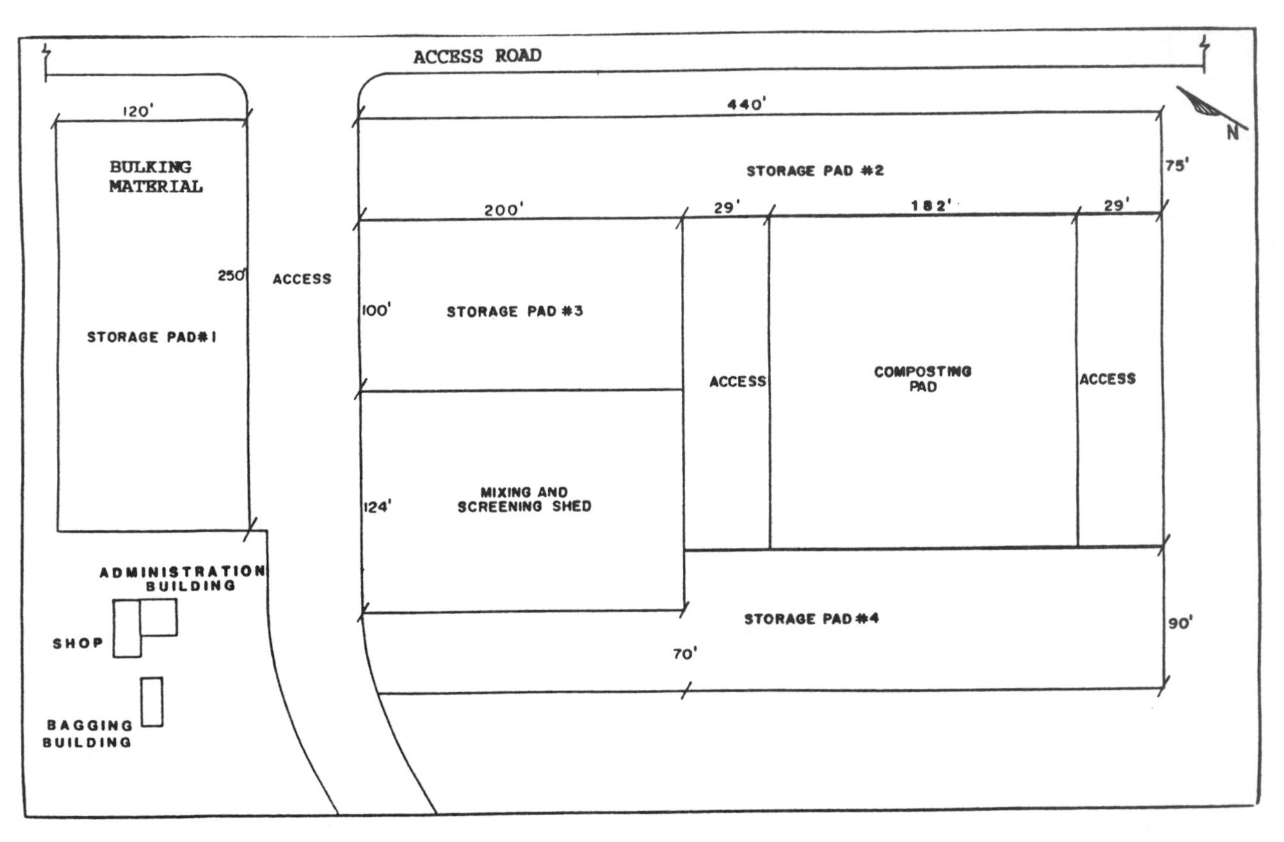

Figure 4.5 ASP model composting facility diagram.

Figure 4.6 Ram-E-Jec trucks used to transport biosolids.

tities using truck and front-end loader volumes as a gauge. (A summary of biosolids and bulking material characteristics is presented later in this text.)

Mixing is performed on the enclosed mixing and screening pad area, a concrete pad measuring 124 feet by 200 feet, for a total of 24,800 square feet. The mixing process, using a front-end loader, takes approximately twenty minutes. As many as three batch truckloads can be mixed at one time under the canopy; when this occurs, maneuvering of mobile equipment is performed outside the canopy.

COMPOSTING

After the biosolids and bulking materials have been mixed, composting piles are constructed using front-end loaders on uncovered concrete pads (see Figure 4.7). The composting pad is a concrete pad with troughs, cast-in-place, for aeration.

Typically, a 1 1/2′ layer of bulking material is placed on the pad first followed by the compost mixture. Dimensions and total area of the pad are 204 feet by 240 feet and 48,960 square feet, respectively, of which 36,312 square feet are dedicated to active composting. This layer of bulking material acts as an air plenum and allows uniform aeration of the entire composting pile. A layer of cured compost is placed on top of the pile to

Figure 4.7 A front-end loader; the workhorse of composting.

insulate the compost material. Piles are usually 185 feet long, 10 feet wide, and 12 to 13 feet high, including the base and cover. Biosolids are composted for a minimum of 26–28 days in accordance with federal and state regulatory requirements. The entire biosolids composting process will be discussed in detail later in this text.

AERATION AND TEMPERATURE CONTROL

As pointed out, the compost pad is equipped with aeration troughs to distribute air to the compost, which is necessary for biological decomposition and to control excessive pile temperatures. Perforated pipe is placed in the permanent trenches to transfer the air from the blowers through the piles using positive pressure. Twenty-four aeration troughs are spaced at 8′ intervals across the composting pad. Each trough is 8 inches wide, 1 foot deep, and 178 feet long.

These aeration troughs are fed by two main header troughs, which, in turn, are fed by eight blowers. Two blowers operate each quarter section of the composting pad. Rated at 1,200 cubic feet per minute (cfm), each blower (3 hp) provides aeration for six troughs. Recently, three variable-speed 2,400 cfm capacity blowers with automatic controls were installed as part of a pilot study (see Figure 4.8). In order to prevent short-circuiting of the air flow, a pile of unscreened compost is placed at the end of each pile.

Controlling temperature in a compost pile is sometimes difficult because compost is not very homogenous. If the compost is not well mixed, the air flow will short-circuit and will most likely result in uneven temperatures within the pile. If temperatures fluctuate widely, pockets of unstabilized compost may result. Compost temperature at the ASP model composting facility is monitored by inserting temperature probes, which are approximately six to seven feet in length, at six places along the compost pile.

When attempting to understand composting process dynamics related to heat and temperature, an important point to consider is the distinction between heat and temperature. As Finstein et al. (1987) point out, heat is measured in units such as BTUs and calories, and temperature in degrees Fahrenheit or Celsius. Heat is not always easy to measure because it is a form of energy. Measuring temperature, on the other hand, is relatively simple since it is the potential driving force for the transfer of energy.

The ASP model facility has tested a temperature feedback blower control system, which uses the average of the three lowest temperature probes as the basis of control. The objective is to maintain the composting process at temperatures between 55°C and 65°C.

Figure 4.8 Aeration blower used to aerate compost piles.

CURING AND SCREENING

When the composting cycle is complete, the compost pile is broken down and moved by front-end loaders to the curing area adjacent to the composting pad. The compost is cured for a period of thirty days with aeration. The aerated curing step increases the total solids concentration in the compost, which improves the efficiency of the screening process.

The compost material is screened after curing using two mobile diesel vibrating deck screens. These screens have a capacity of approximately 60 to 70 cubic yards per hour. Dusting and plugging problems are frequently encountered during screening operations. The existing screens are very maintenance intensive and require frequent repair.

PRODUCT DISTRIBUTION AND MARKETING

The final compost product, Nutri-Value, is distributed and marketed in bulk quantities and as a bagged product. Currently, the bulk market is stronger than the bagged product market; however, bagged product sales have increased steadily in recent years. Bagging equipment is installed on-site and product can be bagged as needed to supply product demand. The product contains approximately 55% total solids and resembles a soil-like, mulching material. With product advertisement and other product marketing efforts, revenues from product sales continue to increase.

REGULATORY CONSIDERATIONS

Regulations concerning the design and construction of composting facilities are important considerations when establishing the preliminary engineering plan for any new composting facility. This section discusses those regulations dealing with facility design, including new Part 503 regulations and state (Virginia) regulations pertaining to final compost product end-use.

FEDERAL REGULATIONS

As stated previously, the U.S. Environmental Protection Agency (EPA) signed new biosolids regulations, 40 CFR Part 503: *Standards for the Use or Disposal of Sewage Sludge* on November 25, 1992. These new regulations were promulgated on February 19, 1993. The following biosolids utilization and disposal practices are covered in these regulations: (1) land application, which includes distribution and marketing; (2) surface disposal; and (3) incineration. General requirements are specified for each option, such as material limits, management practices, monitoring, recordkeeping, and reporting requirements.

The production of compost, such as Nutri-Value, for distribution and marketing is regulated under the land-application procedures. For distribution and marketing, the compost must meet pollutant concentration ceiling (PCC) limits for ten metals as well as annual metal pollutant loading rates if the product is applied for agricultural purposes. For lawn and garden use, the compost must meet pollutant concentration limits specified in 503.13 (b) (3). In addition to metal limitations, application rates cannot exceed the agronomic rate for that soil and crop type. Compost for distribution and marketing must also meet Class A pathogen and vector attraction reduction requirements. Bagged or similarly enclosed compost must carry a label containing the name and address of person responsible for preparing biosolids, a statement prohibiting application except in accordance with instruction on the label or information sheet, and the application rate that will not cause the annual pollutant loadings to be exceeded. Figure 4.9 shows an informational brochure and product label used by Hampton Roads Sanitation District for its compost product Nutri-Green. The compost product label should also include the labeling requirements specified in the 503 regulation (see Figure 4.10).

The original implementation schedule for the 503 regulation, as referenced from the date of promulgation, was as follows:

- Compliance with recordkeeping requirements was required by July 20, 1993.
- Compliance with the 503 regulations in their entirety was required by February 19, 1994.
- If construction was necessary to meet the regulations, compliance was required by February 19, 1995.

STATE REGULATIONS

Regulations governing the treatment, handling, and utilization/disposal of wastewater biosolids vary from state to state. In this text, the regulations applicable to wastewater biosolids treatment, handling, and utilization/disposal in Virginia will be addressed. In Virginia, under the Department of Health, these regulations are entitled Sewage Collection and Treatment Regulations (SCAT).

Along with an incorporation of the essence of the federal part 503 regulations, they also include a Manual of Practice regarding composting facility design considerations. General design information concerning location, odor control, access, and site drainage requirements are given. In addition to site requirements, SCAT includes design considerations for

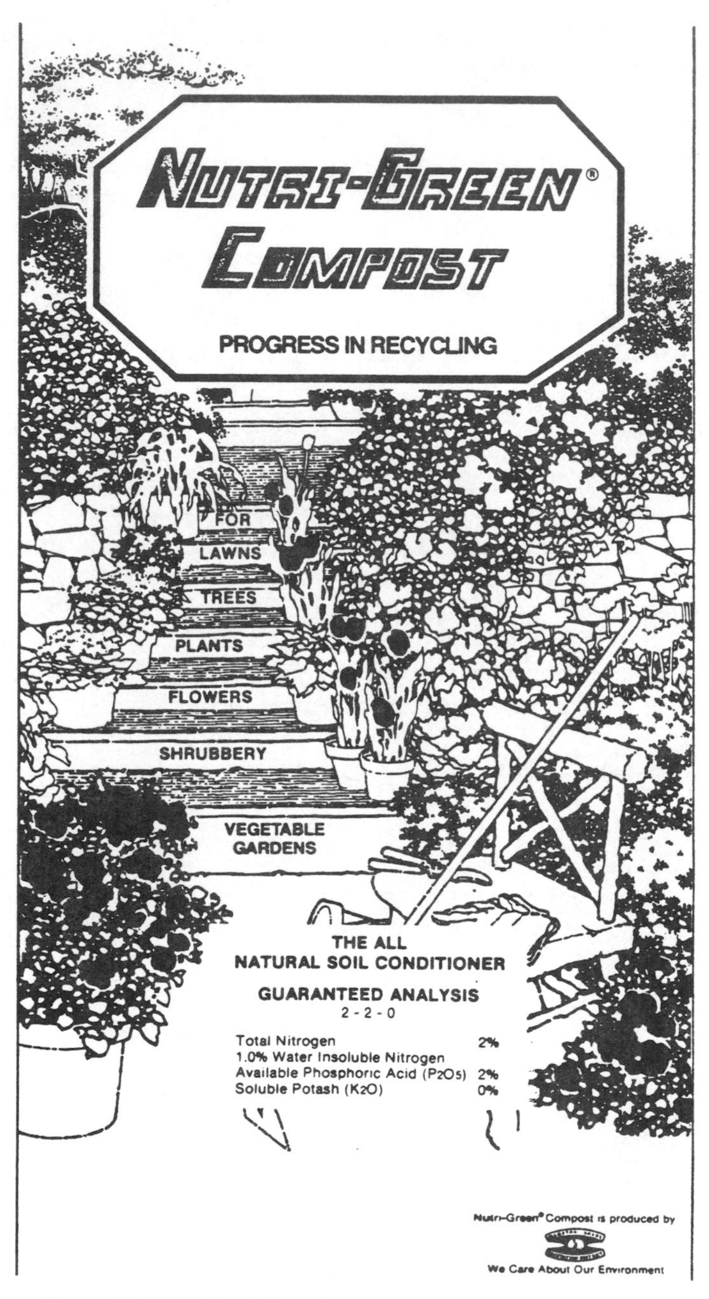

Figure 4.9 HRSD-Nutri-Green information label; used with permission.

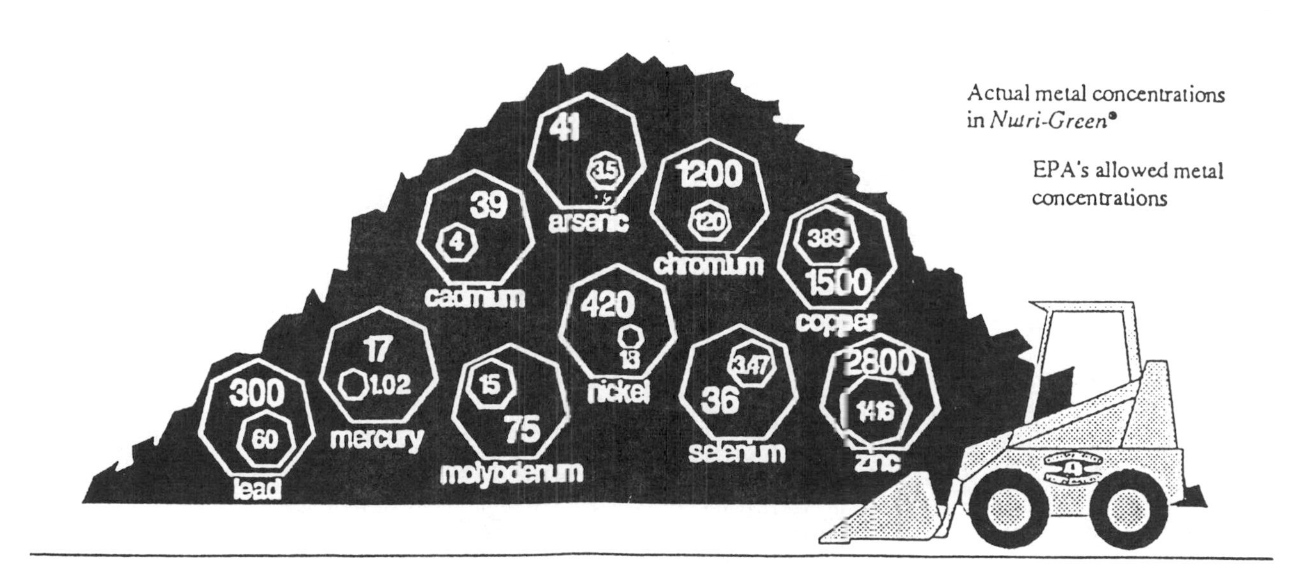

Figure 4.10 Label added to HRSD's Nutri-Green compost product detailing EPA 503 rule required information; used with permission. The information shown on this label predates EPA's recent changes pertaining to chromium and selenium.

biosolids/amendment mixing, equipment selection, process monitoring and aeration, screening, and product storage. Many of the requirements are specified to ensure adequate capacity and redundancy so that composting operations can continue without adverse environmental, health, or nuisance conditions.

ODOR REGULATION AND EMISSION CONTROL

Odor control and emission standards for the ASP model composting facility are governed by the Virginia Air Pollution Control Board, Rules 4–2 and 4–3, respectively. Rule 4–2 is typical of odor control in many states. It applies to any facility in Virginia that emits an odor, other than accidental or infrequent emissions. The rule states that "no owner or other person shall cause or permit to be discharged into the atmosphere from any affected facility any emissions which cause an odor objectionable to individuals or ordinary sensibility."

The Virginia State Air Pollution Control Board determines if an odor emission violates the rule. The board may hold public hearings to hear complaints and/or use other methods such as an odor panel survey to determine violations. If the rule is violated, the facility must use measures approved by the board for economic and feasible control of odorous emissions.

Rule 4–3 provides another method of establishing odor emission control requirements through the use of the emission standard for noncriteria pollutants. This rule regulates the emission of specific noncriteria pollutants. These are pollutants for which no ambient air quality standard has been established. The purpose of the noncriteria pollutant standard is to protect human health.

Uncontrolled odor emission is a nuisance and a sensitive subject with both the public and the politicians who serve the public. Thus, design criteria for the odor control system at any composting facility should be based on state-of-the-art technology. More will be said on this important topic later in the text.

CHAPTER 5

Bulking Material

BULKING MATERIAL

OVERVIEW

JUST prior to delivery of the dewatered biosolids cake to the ASP model composting facility, the biosolids cake has been tested for percent solids concentration, and the results have been recorded in the wastewater treatment plant's daily operating log. The treatment plant then notifies the ASP model facility (via telephone or radio) that a shipment of biosolids has been loaded on the Ram-E-Jec truck and is to be delivered to the composting site. When personnel at the composting site receive notification of biosolids delivery, they begin making predelivery preparations.

The first step taken by compost site personnel immediately prior to biosolids delivery involves moving bulking material (woodchips) from its storage pad to the mixing pad. The bulking agent is spread by a front-end loader over an area of about 40 feet in length and 10 feet in width to a depth of about 8–16 inches. When the truckload of biosolids arrives, the truck backs over and onto the bulking agent bed and slowly dumps a steady measured stream of biosolids cake as it moves along the bulking agent bed. After the truck has emptied its load, a front-end loader quickly moves in and begins the mixing process [see Figure 5.1(a), (b), (c)].

IMPORTANCE OF BULKING AGENTS

A bulking agent is a critical amendment that is added to the biosolids, primarily to reduce the bulk weight (and provide other benefits). It directly affects the composting process and the quality of the final product (Metcalf & Eddy, 1991). The ideal bulking agent has low bulk weight, is readily degradable, and is as dry as possible (Haug, 1980). Bulking agent characteristics such as particle size, moisture content, and absorbency are also important (Corbitt, 1990).

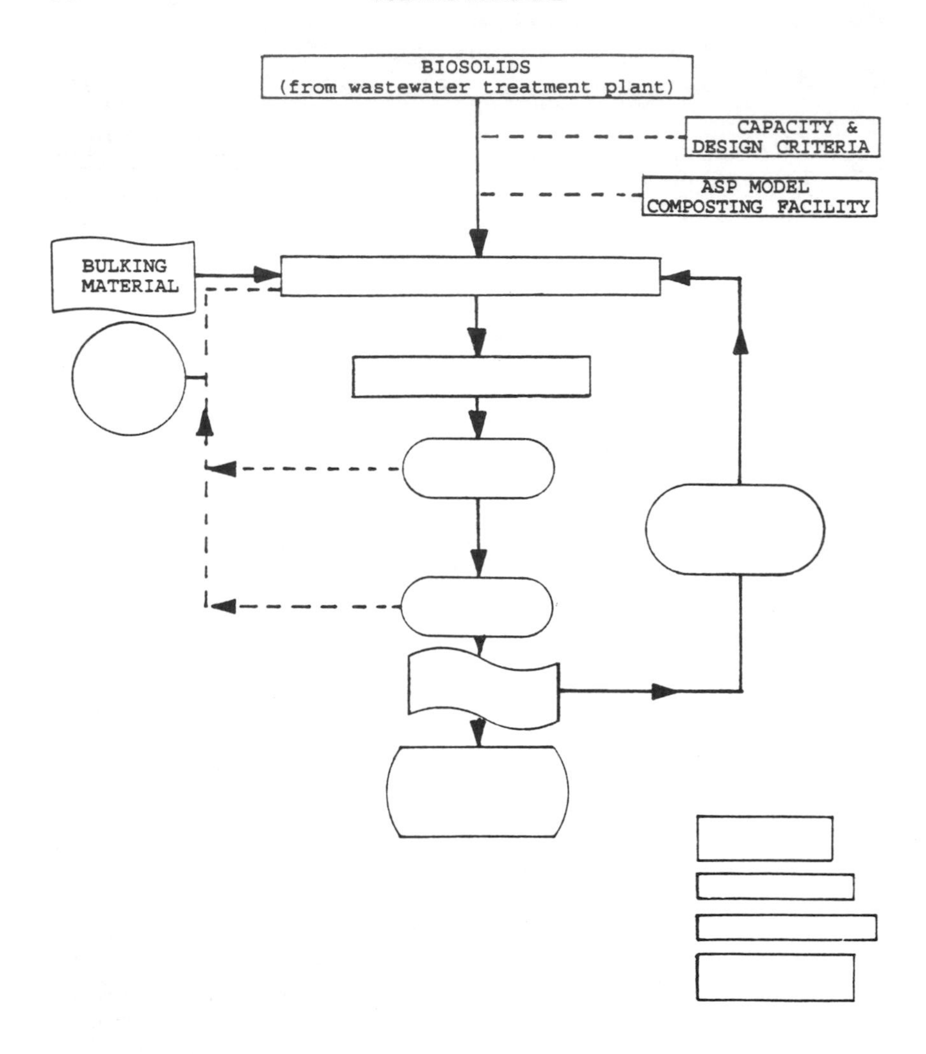

When biosolids cake is at about 22% solids concentration, it is added to the bulking agent that is about 60% solids and mixed or blended. The goal is to attain a mix of 40% solids, which is the minimum solids concentration for this type of composting. This concentration level can be achieved at the ASP compost facility by mixing bulking agent to biosolids in a ratio of 2.5:1 or 3:1 by volume.

Along with adding to the overall percent solids concentration of the biosolids and bulking agent mixture, the bulking agent provides other beneficial functions. Depending upon the bulking agent type, it may provide a source of carbon as a supplemental food source for microorganisms (SSSA, 1994). Additionally, bulking agents increase and help to maintain

(a)

(b)

Figure 5.1 A front-end loader distributing biosolids on bed of woodchips (bulking agent).

(c)

Figure 5.1 (continued) A front-end loader distributing biosolids on bed of woodchips (bulking agent).

porosity; that is, they provide air passages for aeration and, at a minimum of 40% solids concentration, help to prevent leachate drainage (see Figure 5.2).

Figure 5.2 shows that woodchips form a three-dimensional matrix and are in contact with each other; it also shows where biosolids cake can occupy the interstices between the woodchips (Haug, 1980). During the biosolids-bulking agent mixing process it is critical to ensure that enough bulking agent is added. Without the proper ratio of biosolids to bulking agent, it would be impossible for the woodchips to contact each other and form the matrix as shown in Figure 5.2.

It would be difficult for the composting process to progress as designed unless the mixture (biosolids and bulking biosolids) has a consistency of very dry, crumbly bread pudding. If the compost mix is too fluid, the microorganisms that do the work in composting would not be able to obtain enough oxygen. The addition of a bulking agent aids in maintaining structural stability of the static pile. Without the addition of some type of bulking agent, the biosolids, which can have up to 80% moisture content, would be compressed and compacted under its own weight (Vesilind, 1980).

TYPES OF BULKING AGENTS

The type of bulking agent selected for the ASP model composting method is a critical factor for several reasons. According to Shea et al. (1980), these reasons are as follows:

(1) The selected agent must serve to "balance" the mix being composted in terms of moisture and nutrient composition, and must impart free air space and porosity to the mix to ensure uniform air flow.
(2) The selected agent impacts on the quantities of materials handled and, if the replacement cost is such, mandates the recovery by screening to minimize overall net cost.
(3) The selected agent itself may decompose biologically in the process, not only increasing the quantity and cost of makeup but also serving as a substrate for certain microorganisms with undesirable characteristics.
(4) The selected agent may itself be a commodity as well as a byproduct, and as such be subject to price escalation far greater than the current rate of machinery and labor cost (p. 20).

Over the years, different bulking agents have been successfully used in composting operations. Bulking agents include peanut hulls, leaves, bark,

Figure 5.2 Interstices in bulking agent, which are important in providing spaces for air and water.

shredded tires, corn cobs, paper, sawdust, woodchips, and some of the finished compost is often set aside and used later as a bulking agent. In a comparison at one compost facility between using recycled woodchips and recycled compost, the results indicated that for economic and environmental reasons it would be prudent for some facilities to use recycled compost instead of woodchips (Shea ct al., 1980).

In another study where shredded tires were used as the bulking agent in biosolids composting, Higgins et al. (1986) concluded that the increasing consideration given to using this material may be unwarranted. During their study, the researchers found two main problems with using shredded tires: (1) It was difficult to obtain a high-quality material with uniform dimensions; and (2) when shredded rubber was used alone, the biosolids tended to form balls, the interiors of which became anaerobic. This balling tendency lessened considerably when sawdust was added to the rubber-biosolids mix.

The ASP model composting facility uses woodchips as the bulking agent. Woodchips are the bulking agent of choice because they offer many important advantages for composting such as size uniformity, moisture-

TABLE 5.1. Woodchip Specifications.

Wood	All chips are to be produced from freshly cut wood.
Species	Hardwood chips must be manufactured and shipped separately. Loads of mixed species will not be accepted.
Chip size	Will be 1 1/2 to 2″ × 1/4″ thick. All loads will be inspected after unloading.
Foreign material	No metals, glass, plastic, or other foreign substances shall be included among chips.
Rejects	All chips not conforming to these specifications may be rejected by the ASP compost facility.
Chip classification	Chip classification size will be determined by feeding 10 cubic yard from a load through ASP compost facility screen. The amount retained on the 4″ × 0.62″ top deck will determine size of chip. The amount passing the 0.22″ bottom deck (undersize chips) shall not exceed 25% of chips delivered. This could take fifteen to thirty minutes depending on our regular operations.
Clean trucks	All trucks shall be inspected prior to loading to ensure that the truck is free of all foreign matter.
Saw dust	Shall not exceed 1% of the chips delivery. If the content of saw dust is excessively high, it will be rejected.
Bark	No bark will be accepted.
Deliveries	Deliveries will be from 6:30 A.M. until 4:00 P.M. Monday through Friday and Saturday from 7:30 A.M. until 4:00 P.M.

absorption capacity, relative handling ease, and available carbon (Higgins, 1983).

If woodchips and other organic materials are not readily available, what then? DeBertoldi et al. (1980) conducted microbial studies in Italy using various materials as substrates in biosolids composting, including inert recyclables. Their study may provide the answer to the "what-then" question: "it is indeed possible to compost biosolids mixed with inert compounds and the main purpose of using the inert bulking agent is to aid in the aeration and drying of the composting material. Such an agent is particularly useful where the availability of organic mixing material is uncertain" (p. 34).

Assuming woodchips are the bulking agent of choice and that it is readily available, what type of wood should the chips be derived from? The preferred type of woodchip comes from hardwood trees. When soft woodchips are used, too much moisture is absorbed and, because of their structural weakness, they are easily deformed (usually flattened) when compressed. When bulking agent is deformed, the result is a lessening of the space available (collective volume of the interstices is reduced; see Figure 5.2) for air flow (Diaz et al., 1993). Moreover, the screening process and the subsequent recovery of used woodchips is more difficult with soft woodchips than when hard woodchips are used. Table 5.1 shows woodchip specifications for the bulking agent used at the ASP model compost facility. The ASP facility's woodchip suppliers receive a copy of these specifications; when delivered, the woodchips are inspected and if they do not meet the specifications, they are not accepted.

CHAPTER 6

Receiving, Mixing, and Material Handling Requirements

PRELIMINARY INFORMATION

BEFORE discussing mixing, receiving, and material-handling processes/requirements for the ASP model composting facility, it is important to address various factors or elements that affect the composting process.

pH

The closer the pH of the biosolids is to the neutral value of 7, the more efficient the composting process will be. Haug (1980) points out that most raw biosolids have a pH ranging from about 5 to 6.5. Digested biosolids, on the other hand, typically range from 7 to 8 in pH value. The pH value normally drops slightly (to about 5) in the early stages of the composting process because of acid formation. Succeeding microbial populations use these acids as substrates (food) to feed on. As the compost process progresses, the pH level begins to rise and may reach a level of 8.5 (Diaz et al., 1993).

MOISTURE

Moisture content is probably the single most important factor determining the success or failure of the composting process. The profound influence moisture content has on the biosolids composting process can be seen in its interrelationship with aeration. That is, moisture content and oxygen availability are directly related. This interrelationship becomes clear when one remembers that bulking material is added to the biosolids cake to increase the aeration process; that is, to make it easier for oxygen to make contact with the biosolids being composted.

In a properly constructed compost pile, with the correct ratio of bulking material to biosolids, oxygen contact with the biosolids is enhanced

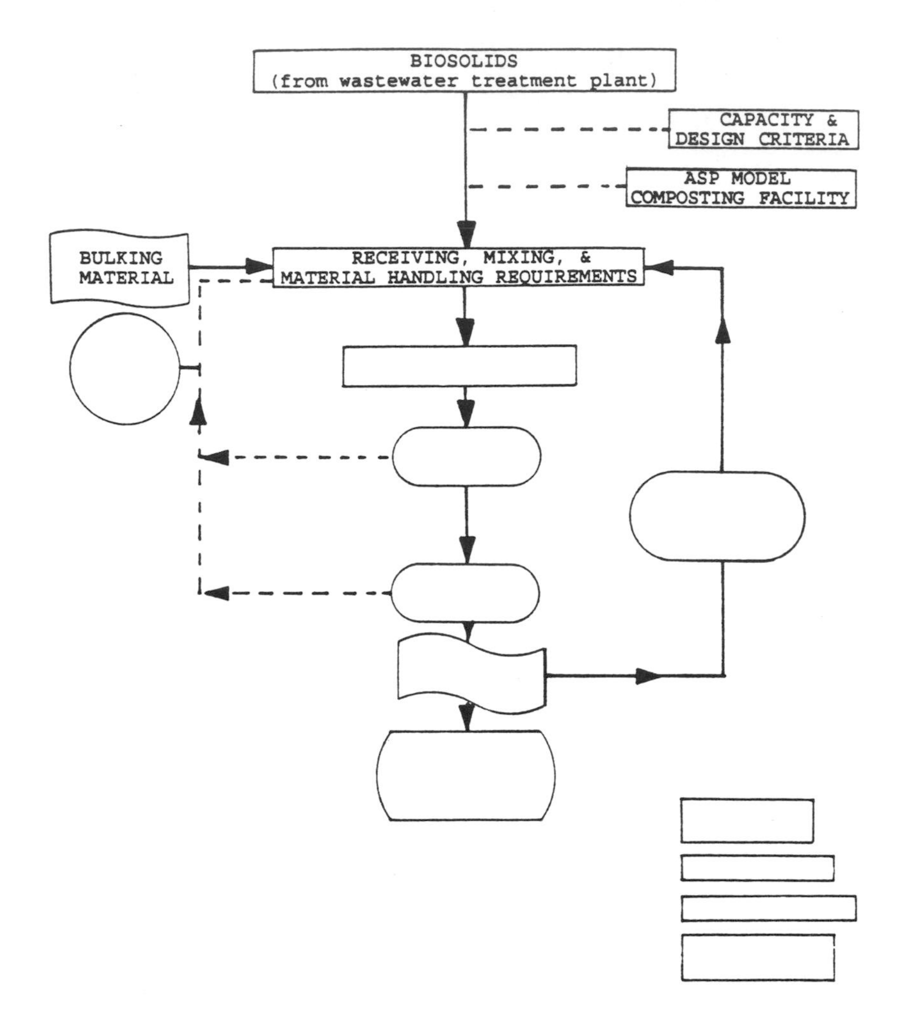

because of the interstices created by the bulking agent, which allow for free flow of oxygen. Too much moisture, however, can block or plug these interstices, making free flow of oxygen more difficult or impossible and creating conditions that enhance anaerobic conditions. This is not to say that some degree of moisture content is not important in the composting process. For example, the biological processes necessary for degradation of the organic material to a finished compost product are dependent on moisture content. But too much moisture content is just as detrimental to the composting process as is too little moisture. The degree of moisture content is measured in the physical unit: % moisture. Typically, the % moisture for bulking agent is greater than 45% and 35% to 65% for the compost mix.

CARBON:NITROGEN (C:N) RATIO

The C:N ratio is an important parameter in composting because it provides a useful indication of the probable rate of decomposition of organic matter. The majority of the nutrients needed to sustain microbial decomposition in composting are readily available in biosolids. Moreover, with regard to the nutritional needs of the microbes active in the composting process, the C:N ratio of the biosolids to be composted is a critical factor. Carbon and nitrogen are used by microbes in obtaining energy through metabolism and in the synthesis of new cellular material (Diaz et al., 1993).

There are those who argue that "carbon availability in compost prepared from biosolids depends on the bulking agents used during the preparation of the compost" (SSSA, 1994, p. 97). Others like Golueke and Diaz (1987) take the view that "woodchips and sawdust should be considered almost entirely as bulking and moisture absorption agents, and very little, if any at all, as sources of carbon" (p. 25). Haug (1980) points out that "during active aerobic growth, living organisms use about 15–30 parts of carbon for each part of nitrogen" (p. 345).

Thus, biosolids mixed with bulking agents with a total C:N ratio from about 25:1 to 30:1 allows the composting process to proceed in an efficient manner. When biosolids with a high nitrogen content of about 11:1 is mixed correctly (in proportion) with a wood bulking agent with a high carbon content of about 700:1, it is possible to obtain a C:N ratio of about 30:1, which is close to optimum for composting biosolids. The key point is that whenever bulking agents are used with biosolids, care must be taken to ensure that the C:N ratio is maintained at the proper level.

TEMPERATURE

Temperature is one of the most important indicators of proper compost performance. Thus, the interior temperatures of compost piles should increase to 50°C or higher within three days. Moreover, it is important to keep in mind that temperatures of 55°C for three days are required by the EPA's 503 rule to kill pathogenic organisms.

OXYGEN LEVELS AND AERATION RATES

Oxygen is necessary for proper development of thermophilic microorganisms and odor abatement. Low oxygen levels could cause odors when the compost pile is broken down. Excessive levels tend to keep temperatures low and slow the rate of composting. Proper control of aeration rates maintains oxygen at proper levels and controls excessive pile temperatures. Oxygen levels and aeration rates and their importance in composting will be discussed in greater detail later.

ROLE OF MICROORGANISMS

When discussing factors or elements that are important to the composting process, the role that microorganisms play cannot be overlooked.

Without the decomposition of the organic matter in biosolids by microorganisms, the composting process would not be possible. This decomposition of the biosolids-compost mixture is accomplished by different types of microorganisms that are active at different times. Microbial activities within the compost pile proceed in a semi-cyclical or sequential manner (see Figure 6.1). This is to say, in the initial stage of the decomposition process bacteria take a commanding position over other types of microbes and process decomposable nutrients such as proteins, carbohydrates, and sugars.

Fungi also play an important role in the decomposition process. While competing with bacteria for food in the earliest stages of the composting process, the fungi begin to predominate as the process goes from its initial wet stages to its drier stage. When the compost begins to dry, bacteria have difficulty in adapting to the drier, low-moisture environment. Fungi have another advantage over bacteria in that they have less need for nitrogen. Also, they are able to feed on cellulose materials, which aids in the composting process. One fungi species that is of concern to compost-site oper-

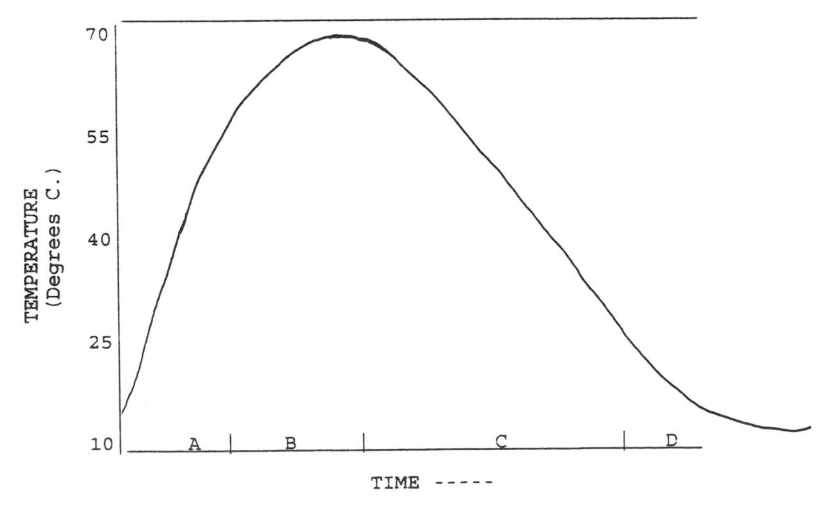

Figure 6.1 Temperature variation with time: phases of microbial activity. A = mesophilic, B = thermophilic, C = cooling, D = maturing. Source: Adaptation of Gray et al. (1971).

ators is the genus *Aspergillus.* (More will be said about the potential health risks and safety concerns of *Aspergillus* later in this text.)

Along with bacteria and fungi, microbes such as rotifers, mites, nematodes, beetles, earthworms, and other macroinvertebrates also play a role in the composting process. By their movements and by feeding on compost materials, these microorganisms contribute to the physical breakdown of the biosolids-compost mixture, which, in turn, helps create more sites for microbial activities to occur.

The cyclical or successive nature of microbial activity within the compost pile begins from the start of the process. For example, when construction of the compost pile is complete, an initial phase of microbial activity begins with mesophilic microorganisms (those that grow best at temperatures ranging from 20°C to 45°C), which dominate the pile. By using the available oxygen to transform carbon from the biosolids, they are able to obtain the energy they need. During this oxygen-carbon-to-energy process, the microorganisms produce water and carbon dioxide. They also produce heat, which is important (critical) to the composting process.

Remember: EPA's 503 rule requires maintenance of at least 55°C temperatures for at least three days during the composting process. More will be said about the 55°C factor. For now, the temperature in the compost pile is increasing from its initial level and has not yet reached 45°C. The subsequent relationship with increasing temperatures and the composting process is best understood after comparing the activities that take place in a nonaerated static pile and an aerated static pile composting operation.

Nonaerated Static Pile

In a compost pile without forced aeration, the microbes continue to do their work and to build up heat that will be trapped inside a well-insulated pile. The temperature will continue to rise until the mesophilic organisms can no longer tolerate the increased temperatures. Thus, when the temperature reaches 45°C or greater, the mesophiles die off or enter dormancy (they will reactivate as the cycle continues and temperatures return to their tolerance range).

The activities of the mesophilic microbes increase the temperature within the biosolids-compost pile in a few days. When the 45°C temperature level is reached, conditions are right for the thermophilic microbes to activate and take over the biosolids-compost degradation process (Singleton & Sainsbury, 1994).

Thermophile microorganisms prefer temperatures between 45 and 70°C. They quickly become active, consume material, reproduce, and replace the mesophiles. Because the rising temperatures also increase the thermophiles' rate of metabolism, the heat-generation process increases to higher levels. This increase in temperature is important because the heat generated must be hot enough to destroy any pathogenic microorganisms. When the temperature reaches 55°C (131°F) or greater, it should be maintained at this level for at least 72 hours. (As stated earlier, this is also a requirement under EPA's 503 rule.)

The composting process continues with the thermophiles decomposing the biosolids-compost materials until the nutrient and energy sources are depleted. When depletion occurs, the thermophiles die off and the temperature of the pile decreases. At this point, the mesophiles reactivate and generally dominate the decomposition process once again until all energy sources are depleted. Thus, the microbial succession cycle and this phase of the composting process is complete.

Aerated Static Pile

During the preceding discussion, the microbial succession cycle and subsequent degradation of biosolids-derived compost were accomplished in the natural mode; that is, without forced aeration. In the ASP model, forced aeration is an essential part of the process.

Using forced aeration in static-pile composting provides several advantages over nonaeration. In the first place, by using forced aeration, the static pile can remain static. That is, it does not have to be routinely "turned over" to facilitate better aeration. Another advantage of forced aeration is that it basically gives a "kick start" to the biosolids-compost degradation process by ensuring that decomposition proceeds at a high rate. Moreover, forced aeration produces aerobic conditions, which prevents the process (pile) from going septic. Yet another advantage of forced over nonforced aeration is that in forced aeration control of temperatures from 55°C to the 70–75°C temperature range is easier.

Generally, in the initial stages of the process the aeration blowers are activated by timers that start the blowers for predetermined periods to create aerobic conditions that are conducive to microbial growth and activity. As the microbes multiply and consume organics, the temperatures rise (usually within a few days) to the 55°C level. When this happens, in the ASP model, for example, the aeration blowers are automatically shifted from manual to automatic.

As pointed out earlier, in automatic mode, the aeration blowers run intermittently or continuously to maintain composting temperatures between 56 and 75°C. When temperatures start to rise too quickly, these automatic blowers go to 100% operation (via a feedback system), providing forced air to cool down the pile. This is important because if the compost pile is allowed to increase in temperature above the 75°C setpoint, spontaneous combustion might occur and fire will result. When this happens, the compost pile can quickly turn into a smoldering, smelly mess that could eventually turn into an uncontrolled bonfire.

Maintaining this 55–75°C temperature range is important for another reason. When temperatures reach the higher levels and when conditions allow, copious amounts of steam might be produced. This steam has the tendency (depending on wind conditions) to hover over the pile and condense. Such condensation can turn a drying compost pile into a mess that can only be described by compost facility operators who have personally faced this nasty problem.

From the previous discussion it is apparent that temperature control is important in the aerated static pile composting process. As previously mentioned, the ASP model composting facility uses a temperature feedback system that operates when the system is set up in automatic mode. This feedback system attempts to maintain optimum pile temperatures. Electronic sensors, such as the thermocouples shown in Figures 6.2 and 6.3, switch the blowers on and off when the temperature exceeds or falls below predetermined levels.

It should be pointed out that there is also a disadvantage associated with using forced aeration in composting process. In the positive pressure mode, blowers push air into the pile, which causes the forced air to be vented over the pile's entire surface area. While it is true that the air has

Figure 6.2 Thermocouples that control aeration blowers for composting and curing/drying piles and their associated wiring.

Figure 6.3 RTD mounted in T of aeration piping. The RTD feeds a signal through wiring to an associated thermocouple.

to go somewhere, it is also true that the forced air carries and delivers odors to areas where they may become a nuisance.

As stated previously, in composting biosolids it is important to ensure that the proper mixture of bulking agent to biosolids is accomplished. Proper mixing is critical because of the importance of creating a porous biosolid/bulking agent mixture that allows air to flow freely throughout the compost pile, which aids in achieving optimal composting conditions.

Several options are available to compost-facility designers in choosing the mixing equipment for their operation. In the ASP model, front-end loaders are used for mixing. In other compost facilities, mixing systems may consist of any of several different types of mechanical mixing devices, including the Van Dale, rotary drum, and McLanahan plug mill mixers (Higgins et al., 1981).

Along with discussing the front-end loader mixing process used at the ASP model composting facility, the following sections will present a scenario whereby a mechanical mixing system is utilized in a compost facility setting. Moreover, front-end loader and mechanical mixing will be compared and evaluated.

MATERIAL QUANTITIES AND CHARACTERISTICS

When attempting to determine the best method to use in the mixing part of the composting process, it is important to compare and contrast the alternatives. Two alternatives are presented here: front-end loader and mechanical mixing systems.

In comparing the front-end loader and mechanical mixing alternatives it is important to point out the material quantities and characteristics used in making the comparison. For example, woodchips will be used as the base upon which the composting piles are built. To the extent available, recycled woodchips will be used in mix preparation. New woodchips will augment the mix as necessary to achieve a minimum compost-mix solids concentration of 40%. The mix ratio for this comparison is 2.5:1 of woodchips to biosolids.

The materials handling systems described here will be sized based on a peak weekly biosolids production rate of 185.5 dry tons per week (dtpw). The mechanical mixing system will mix biosolids at an hourly rate. Front-end loader mixing is based on mixing biosolids on a truckload basis. Mixing will be conducted for 6.5 hours each day for a five-day work week. Table 6.1 shows the material design criteria for the mixing system for biosolids, recycled woodchips, and mixed material. Volume requirements are provided for both an hourly and a per-load rate.

BIOSOLIDS RECEIVING AND HANDLING

For illustrative purposes the following scenario is presented. Biosolids will be delivered to the composting facility from two wastewater treatment plants in the region. Based on the peak week biosolids production rate of 185.5 dtpw, four truckloads of biosolids will be hauled from one plant and six truckloads from the other. Each truckload will contain approximately 24 cubic yards of biosolids, although the trucks delivering biosolids at the end of each day may contain less. The method of biosolids receiving and handling depends on whether a mechanical or front-end loader mixing system is selected.

MECHANICAL MIXING ALTERNATIVE

A mechanical mixing system requires mechanical receiving equipment. Biosolids are unloaded from the truck directly into below-grade live bottom bins that store and meter the biosolids to the mixing system. The live

TABLE 6.1. Material Handling Design Criteria—Mixing System.

Criteria	Units	Peak Week 2.5:1 Ratio
Biosolids Criteria		
Feed rate	dtpw	185.5
Bulk density	lb/cubic yard (cy)	1,350
Total solids	%	25
Volume		
Weekly	cy/week	1,099
Daily (5-day)	cy/day	220
Mechanical Mixing		
Hourly (6.5 hours/day)	cy/hour	34
Front-end loader mixing		
Load	cy/load	24
Recycled Woodchip Criteria		
Bulk density	lb/cy	750
Total solids	%	55
Volume		
Weekly	cy/week	2,748
Daily (5-day)	cy/day	550
Mechanical mixing		
Hourly (6.5 hours/day)	cy/hour	85
Front-end loader mixing		
Load	cy/load	60
Initial Mix Criteria		
Bulk density	lb/cy	900
Total solids	%	43.7
Volume		
Weekly	cy/week	3,939
Daily (5-day)	cy/day	788
Mechanical mixing		
Hourly (6.5 hours/day)	cy/hour	121
Front-end loader mixing		
Load	cy/load	86

Source: HRSD/Black & Veatch Engineering Study; used with permission (1993).

bottom bins consist of vertical-sided fabricated steel bins with parallel screw conveyors that form the entire bottom of the bin and a variable-speed discharge screw conveyor. The bottom screw conveyors prevent bridging of material and feed the material to the discharge screw conveyor, which is located beneath and perpendicular to the bottom conveyors. An electrically operated steel slide gate is provided in a chute from the discharge conveyor. In order to provide adequate odor control, the biosolids receiving bins are equipped with power-operated doors. The bins are located below-grade in an outdoor pit. Biosolids trucks discharge directly into the receiving bins.

Figure 6.4 shows the layout for the mechanical mixing system. Each receiving bin is sized to receive one 24-cubic yard truckload of dewatered biosolids. Two 30-cubic yard biosolids receiving bins (RB-1 and RB-2) will be provided for system redundancy and to accommodate two trucks simultaneously. The discharge screw conveyors transport biosolids to a level belt conveyor (BC-1), which transfers to an inclined belt conveyor (BC-2) to life the biosolids above grade. Space is provided to add a third receiving bin and parallel belt conveyors for future expansion.

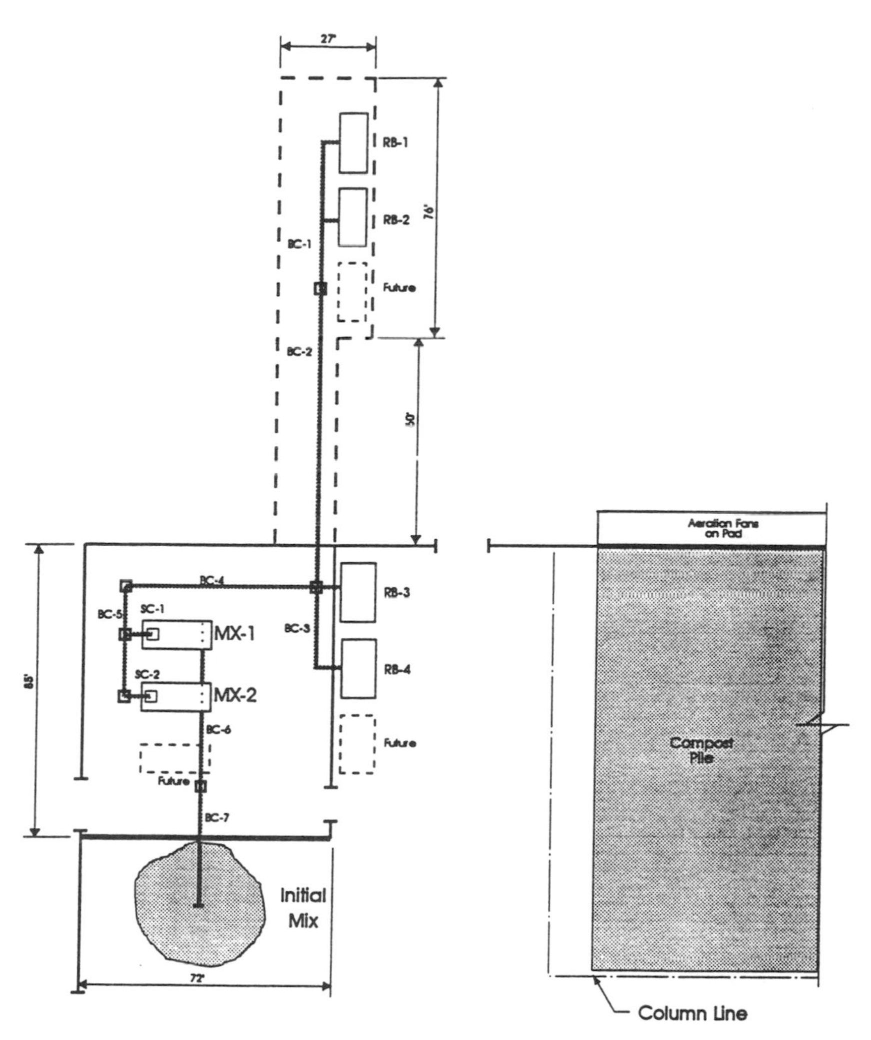

Figure 6.4 Layout for mechanical mixing system. Source: HRSD and Black & Veatch (1993); used with permission.

Mechanical Mixing Equipment and Operation

The mechanical mixing alternative consists of two redundant pug mill mixers equipped with paddles attached to dual shafts. The rotating shafts, which create a shearing effect as the material is mixed, produce a homogenous mixture with little balling of biosolids. Mixing retention time at design capacity is generally one minute. The paddles' shearing action may accelerate woodchip degradation. The intensity of the mix is affected by the speed of the mixer, and variable-speed drivers are provided.

The mechanical mixing alternative uses automated equipment to meter, convey, and mix biosolids and woodchips. Figure 6.4 shows the equipment layout for the mechanical mixing alternative. Truckloads of biosolids are discharged directly into the below-grade outdoor receiving bins (RB-1, 2). Belt conveyors collect the biosolids (BC-1) and lift them (BC-2) from the receiving pit. The biosolids conveyor (BC-2) enters the mixing area above grade and discharges biosolids into a common chute, with woodchips conveyed (BC-3) from the woodchip receiving bins (RB-3, 4). The combined woodchips and biosolids are transported on an inclined belt conveyor (BC-4), which discharges onto a level belt distribution conveyor (BC-5) over the top of the mixers. Plows are provided to discharge material from the belt into short screw conveyors (SC-1, 2) that feed each mixer.

The compost mix discharged from the bottom of the mixers is collected on a common level conveyor (BC-6) and transported on an inclined conveyor (BC-7) to a temporary storage area outside of the mixing area. This stockpile area is able to store at least one hour of initial mix generated at peak week conditions. The compost mix is transported from this temporary storage pile to the active face of the composting pile using a front-end loader.

To accommodate future increases in capacity requirements, space is provided for an additional biosolids receiving bin, woodchip receiving bin, and pug mill mixer. Duplicate conveyors will be required for BC-1, 2, 3, 4 and 5 for two mixers to operate simultaneously. The layout shown in Figure 6.4 provides adequate space for the conveyors required to double the proposed initial design capacity. The enclosed mixing area has approximate dimensions of 85 feet by 72 feet (6,120 sf). The outdoor biosolids receiving bin conveyor is approximately 136 feet long with widths varying between 27 feet and 17 feet.

Air from the biosolids receiving bins and the mechanical mixers is collected and scrubbed to control odors. In addition, the receiving bins have doors that can be closed to contain idors.

FRONT-END LOADER MIXING ALTERNATIVE

Mixing operations can be accomplished effectively using front-end

(a)

(b)

Figure 6.5 (a) Mixing operation outdoors and (b) mixing operation performed under cover.

loaders. This process begins with weighing the biosolids trucks to determine the wet weight of material to assist in proportioning bulking material. Bulk density is periodically measured to correlate wet weight with the volume of biosolids. Biosolids are discharged directly onto a bed of bulking material and mixed using front-end loaders [see Figures 6.5(a) and (b)].

Biosolids receiving and mixing is performed inside an enclosed building. Figure 6.6 shows the layout, with approximately 25,500 square feet of area provided. This area is sufficient to concurrently mix two truckloads of biosolids and to store temporarily one truckload of biosolids.

Front-End Loader Mixing Equipment and Operation

Equipment used in this alternative consists of front-end loaders and a truck weigh scale. As stated earlier, biosolids trucks are weighed to determine the wet weight of material to assist in determining the appropriate mix ratio. Front-end loaders are used to measure the volume of woodchips required. Biosolids are directly discharged from the trucks onto a 8–18-inch base of new and recycled bulking material [see Figure 6.5(a)]. Additional bulking material is added as necessary to achieve the appropriate mix ratio.

Front-end loaders are used to mix the biosolids and bulking material until a consistent mix is achieved. Mixing one truck load of biosolids with a front-end loader requires from fifteen to twenty minutes. A front-end loader is required for approximately 3.5 hours, or for 55% of each 6.5-hour work day, to mix ten truckloads of biosolids. After mixing is complete, the front-end loaders are used to transport the mix to the active face of the composting pile. The 3.5 hours does not account for the time required to prepare the bulking material base or to build the composting pile.

The active mixing area or pad (see Figure 6.6) is large enough for two truckloads of biosolids to be mixed simultaneously. The biosolids trucks back into the mixing area through large access doors and discharge the biosolids directly onto the bulking material base. A 20-foot buffer area around the perimeter of the mixing area protects the building walls. 25,500 square feet of mixing area is required for front-end loader mixing.

Odors are generally contained if front-end loader mixing is conducted indoors (when mixing occurs outdoors, odor generation is almost always a problem). In indoor mixing, building ventilation air from the mixing area is diluted for odor reduction and discharged with high-velocity dispersion fans. Dilution and dispersion of mixing process air does not provide the degree of odor control that chemical scrubbing achieves with the mechanical mixing alternative. If chemical scrubbing of mixing process air is necessary for the front-end loader mixing option, wall partitions would be required to separate the mixing area from the composting/curing area. However, addition of walls to the layout would reduce the amount of active mixing area, because space required for front-end loader access to the composting pile is also used for mixing. In addition, a separate scrubbing system is required to treat the mixing process air.

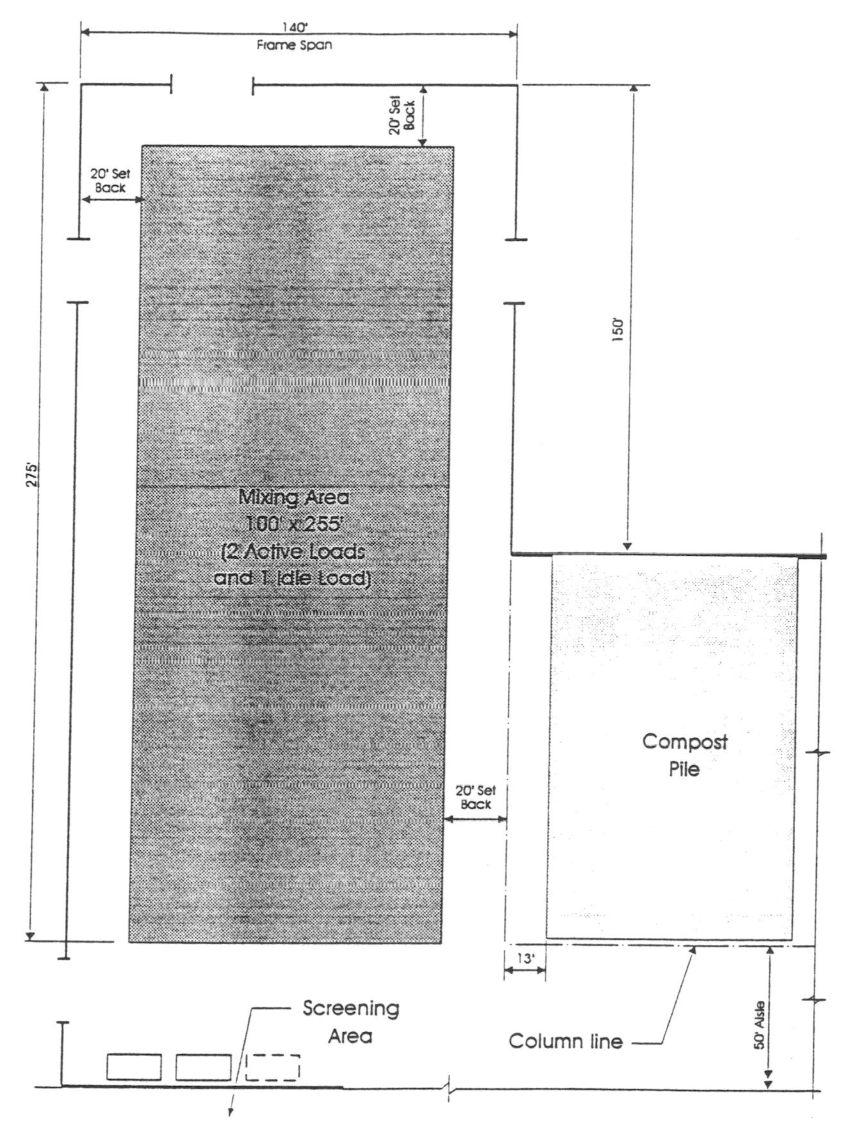

Figure 6.6 The layout of the site used for the front-end loader mixing alternative discussed in this text. Source: HRSD and Black & Veatch (1993); used with permission.

COMPARISON OF ALTERNATIVES

The mechanical and front-end loader mixing alternatives are compared in Table 6.1 based on area requirements, odor control, and system cost.

Area requirements for the two alternatives are compared based on mixing area and additional building area. The front-end loader mixing alternative requires almost four times as much active mixing area as the mechanical mixing alternative.

The mechanical and front-end loader mixing alternatives result in different levels of odor control. Generally, a thorough mix of biosolids and bulking material has a homogenous porosity and produces less odors. Air flow is poor in areas of the pile with insufficient porosity, which could result in anoxic conditions. Odors are produced when sections of the pile are anoxic. Balling in the compost mixture and some types of bulking materials affect porosity. These conditions are more likely to occur during front-end loader mixing and have been observed at other facilities using this mixing system. Mechanical mixers generally produce a more uniform mixture than front-end loaders, which results in less odor. *Note:* It is important to keep in mind that a lack of uniformity in mix reduces the desirability of the final compost product (Higgins et al., 1981).

In addition, approximately 800 cfm of air from the biosolids receiving bins and the mechanical mixers is collected and scrubbed. The receiving bins also have doors that can be closed to contain odors. Air within the front-end loader mixing area is diluted and dispersed rather than scrubbed. For these reasons, the mechanical mixing system is expected to produce less significant odors than the front-end loader mixing system.

The differential cost between the two mixing alternatives is presented in Table 6.2. Capital costs are summarized for each alternative. As shown the mechanical mixing alternative is equipment intensive and requires a basement area for the receiving bins. The front-end loader mixing alternative requires the compost process building to be extended from the baseline. Front-end loader mixing requires a dedicated front-end loader for 55% of each day to handle mixing under design conditions. Partial use of a front-end loader has been included as a capital cost for this option. Front-end loaders are used to manage the bulking material required for both mixing alternatives; therefore, associated costs apply to both alternatives and are not included in this analysis. Operational costs are included for the labor associated with front-end loader mixing. Labor is given as a present worth cost determined over a ten–year period. The mechanical mixing alternative would be approximately 30% more expensive than the front-end loader mixing alternative.

The mechanical mixing system provides an automated mixing system that produces a thorough initial mix. Odors are controlled to a greater ex-

TABLE 6.2. Comparison of Mixing Alternatives.

Line Item	Mechanical Mixing	Front-End Loader Mixing
Area Requirements		
Mixing area	6,120 sq ft	25,500 sq ft
Additional building area	none	21,000 sq ft
Odor Control		
Potential for odors	Odors are controlled; biosolids receiving bins covered and air scrubbed; mechanical mixers produce a better mix with less potential for anoxic zones in compost pile and fewer resulting odors.	Odors will be more significant than mechanical mixing; mixing will be conducted indoors but odors are not scrubbed; biosolids balling may occur and could create anoxic zones and odors in compost pile.
Odor control measures	Scrub air from biosolids receiving bins and mechanical mixers.	Mixing will be conducted indoors; dilution and dispersion.

TABLE 6.2. (continued).

Line Item	Mechanical Mixing		Front-End Loader Mixing	
System Costs				
Capital costs				
Building area	none	$0	21,000 sf	$411,000
Mechanical mixers	2 each	$46,000	none	$0
Receiving bins	4 each	$600,000	none	$0
Belt conveyors	340 LF	$170,000	none	$0
Screw conveyors	20 LF	$16,000	none	$0
Receiving bin				
Conveyor pit	3,072 SF	$135,000	none	$0
Odor control	800 CFM	$40,000	none	$0
Front-end loaders—Mixing	none	$0	55% op day	$138,000
Operating Costs				
Labor for mixing—FEL operation (10 years)	none	$0	3.5 hr/dy	$150,000
Total differential cost	—	$1,007,000	—	$699,000
Net differential cost			$308,000	

Source: HRSD/Black & Veatch Engineering Study, 1993; used with permission.

tent since the bulking material is well coated (with biosolids) and anoxic conditions should be minimal in the compost pile. Although there are several advantages to using the mechanical mixing system alternative, for the purposes of this study the front-end loader mixing system will be used because of differential cost savings as noted in Table 6.2 and other cost savings to be gained through use of equipment already purchased, such as four front-end loaders. At the present time, the front-end loader mixing system is the predominant system in use in biosolids-derived composting operations.

STANDARD OPERATING PROCEDURE (SOP) FOR MIXING: ASP MODEL

To help ensure the proper completion of any work task, it is important that workers know the correct procedure to be followed. Written guidelines that provide step-by-step instructions on how to perform the task are important. Such instructions, which must be sequential and clearly written, serve many purposes. In addition to providing a step-by-step protocol for correct system operation, written procedures assist (1) in accomplishing reliable and efficient system operation; (2) in familiarizing new personnel with system operation; (3) in serving as a reference guide to other organizational manuals; and (4) in ensuring safe working conditions.

SOP FOR MIXING OPERATION: ASP MODEL

Theory of Operation

Good mixing is an essential part of the composting operation. In this process stage, biosolids is mixed with a bulking agent in order to:

(1) Adjust the moisture content of the biosolids bulking agent mixture.
(2) Provide an additional source of carbon.
(3) Increase porosity of the biosolids, thereby allowing passage of air through the pile.
(4) Provide structural stability for the pile (acts as the superstructure of the pile).

Examples of materials that can be used for bulking agent include: woodchips, wood shavings, peanut hulls, sawdust leaves, bark, garbage, paper, straw, plastic, tires, and recycled dried compost. (The bulking agent used at the ASP model composting facility consists of fresh or recycled woodchips.)

Care must be taken to ensure that bulking agent characteristics (i.e., heavy metal content) do not affect the overall quality of the compost.

The quantity of bulking agent necessary for proper mixing depends on the moisture content of both the biosolids and the bulking agent. Generally, the wetter the biosolids, the more bulking agent required. The bulking-agent-to-biosolids ratio is determined on a case-by-case (load-by-load) basis for each pile.

Mixing Procedure

(1) Determine the volume of biosolids to be delivered.

(2) Determine % of solids concentration of delivered biosolids.

(3) Determine volume of bulking agent necessary to achieve a bulking agent biosolids mixture of 50–60% moisture (40–50% TS). The volume of bulking agent will depend on the percent TS of biosolids, and the type and the percent TS of bulking agent to be used. The ratio of mixing bulking materials with biosolids is normally set at 2.5:1, depending on conditions. To make an approximate determination of the ratio of woodchips to biosolids, see Figure 6.7. A more accurate determination of volumetric mixing ratios can be made using the examples presented by Haug (1980) in his test: *Compost Engineering.*

(4) Prepare a bed of the bulking agent sufficient to handle incoming volume of biosolids.

(5) Discharge biosolids from the trailer directly onto the prepared bulking agent bed or place biosolids on the bed with the front-end loader.

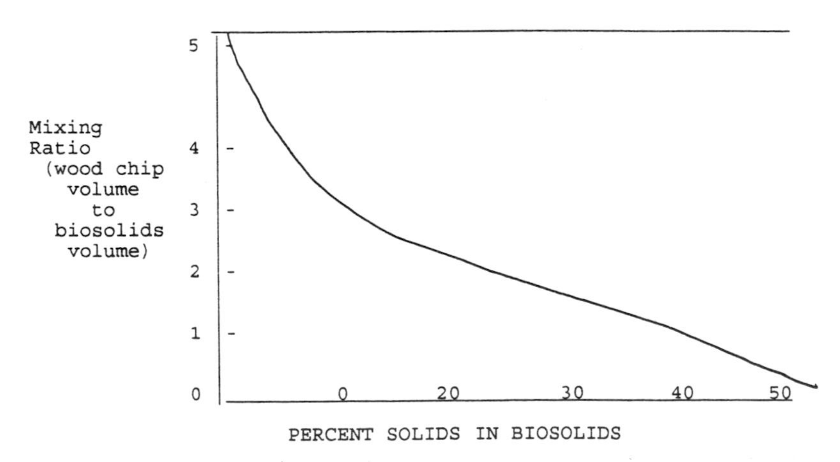

Figure 6.7 Effect of biosolids content on the ratio of woodchips to biosolids by volume. This curve is site specific for one compost operation. It will shift, depending on the relative volatility and solids content of the woodchips and biosolids. Source: Adaptation from Haug, R. T (1980).

(6) Mix biosolids and bulking agent. Mixing is to be done with a front-end loader.
(7) Biosolids can be mixed at:
- mixing and screening pad/shed
- composting and mixing and traffic area(s)
- other outside pads

Note: Proper mixing is essential in composting. Operators must be careful to take sufficient time to ensure an adequate mix. The compost mix should not contain large lumps of biosolids, if so, a slow rate of composting will occur and the compost mix might fail to reach the required temperatures. It is desirable to mix under cover during rain.

CHAPTER 7

Composting: Aerated Static Pile

THE main purpose of this chapter is to describe the aerobic operation and process that (1) is used in the biooxidation of the organic matter (composting) and (2) is used to control the aeration process. To make the study of this operation and process more meaningful, the first section of this chapter is devoted to a discussion of aeration. Because control of the composting aeration process is a fundamental part of the composting facility design, a separate section is devoted to the aeration control process.

AERATION FOR COMPOSTING AND CURING

Aeration is an important process control parameter in the aerated static pile composting system. Air is necessary to supply oxygen for biological degradation of organic solids in the biosolids and woodchips (ASP model). Aeration is also needed for the removal of heat generated by the biological activity in the compost pile and excess moisture from the compost mix. Fans ensure that sufficient quantities of air are supplied to meet composting process requirements and to provide the flexibility necessary for optimizing operations.

PRIMARY REQUIREMENTS FOR AERATING THE COMPOSTING AND CURING PILES

Oxidation Air

The composting process requires oxygen to support aerobic biological oxidation of degradable organics in the biosolids and woodchips. Stoichiometric requirements for oxygen are related to the extent of organic

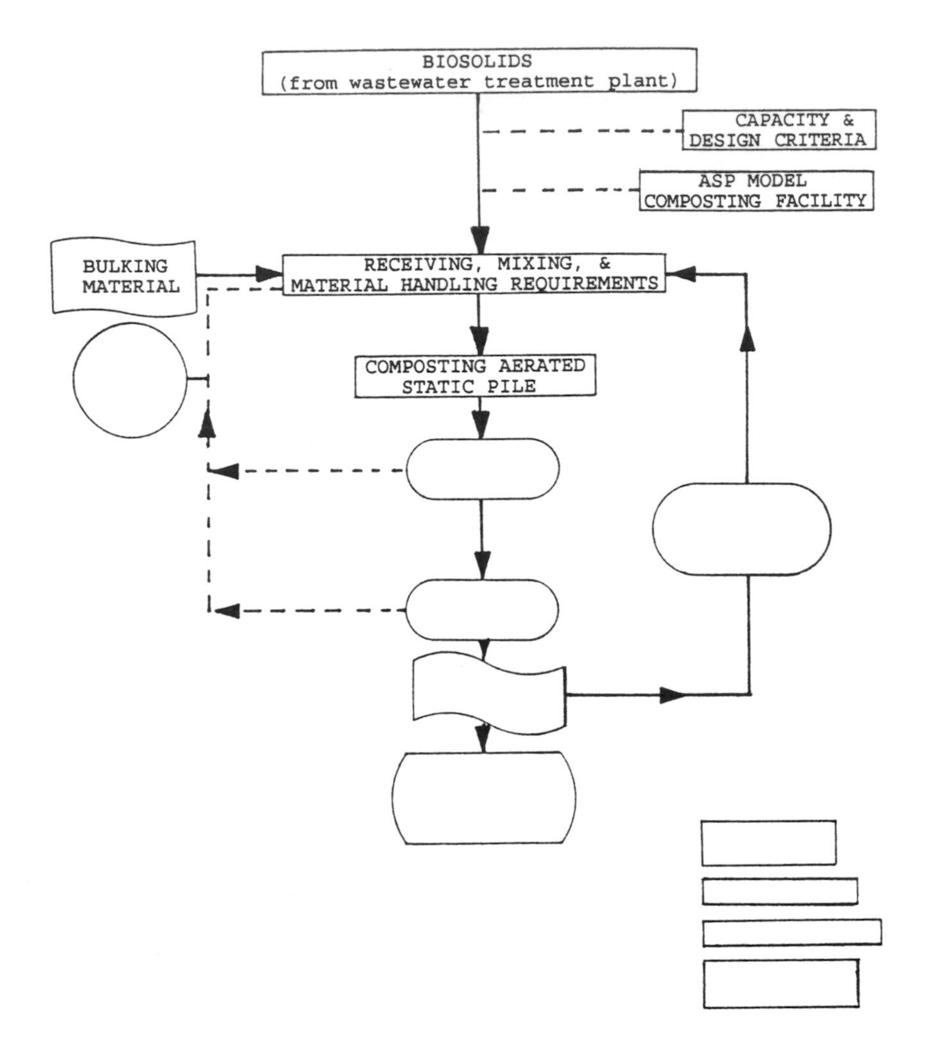

solids degradation expected during the composting cycle time. Haug (1986) estimates oxygen requirements for organic solids degradation by relating the biodegradable organic solids to an organic molecule with the relative proportions of carbon, hydrogen, oxygen, and nitrogen typically found in wastewater biosolids. Minimum stoichiometric oxygen requirements are typically increased by at least 30% to account for oxygen transfer from the air to the biological organisms. The volatile solids fractions and oxygen demand of the biosolids and amendment affect the calculation of minimum air requirements for organic solids oxidation.

Heat Removal and Temperature Control

The biological oxidation process for composting biosolids is an exothermic reaction. The heat given off by the composting process can raise the temperature of the compost pile high enough to destroy the organisms responsible for biodegradation. Therefore, the compost pile cells are aerated to control the temperature by removing excess heat to maintain optimum temperature for organic solids degradation and pathogen reduction. Optimum temperatures are typically between 50 and 60°C (112 and 140°F). Using summer ambient air conditions, aeration requirements for heat removal can be calculated.

Moisture Removal

When the temperature increases, the quantity of moisture in saturated air increases. Air is required for the composting process to remove water that is present in the mix and produced by the oxidation of organic solids. The quantity of air required for moisture removal is calculated based on the desired moisture content for the compost product and the psychometric properties of the ambient air supply. Air requirements for moisture removal are calculated from summer ambient air conditions and required final compost characteristics.

Peaking Air

The rate of organic oxidation, and therefore the heat release, can vary greatly for the composting process. Standard procedure calls for sizing fans to provide sufficient air for periods when peak air demand exceeds average air requirements predicted for the composting process (Haug, 1980). If sufficient aeration capacity is not provided to meet peak requirements for heat or moisture removal, temperature limits for the process may be exceeded. Peaking air rates are typically 1.9 times the average aeration rate for heat removal.

Curing Air

Curing piles are aerated primarily for moisture removal to meet final product moisture requirements and to keep odors from building up in the compost pile as biological activity is dissipating. Final product moisture requirements and summer ambient conditions are used to determine air requirements for moisture removal for the curing process.

PROCESS DESIGN CONSIDERATIONS FOR ASP MODEL AIR REQUIREMENTS

The process design considerations used to determine the air requirements for both the composting and the curing piles for the ASP model composting facility are presented in Table 7.1. These air quantities were used as the basis for determining fan capacities for the aeration system.

AERATION PROCESS CONTROL

Haug (1986) notes that there are "a number of control strategies which have been used in practice to regulate the aeration rate" (p. 57). The key point to remember is that after the minimum, average, and peak aeration rates have been determined, the designer must select aeration supply and control systems that will operate over the full range of flow conditions.

ASP Model Aeration System

Aeration fans and aeration rates can be controlled at the ASP model composting facility. Aeration fan and rate control is important not only for controlling the composting process, but also for controlling temperatures. Temperatures are monitored manually and automatically. The ASP model's aeration process control and temperature monitoring system will be discussed here along with alternative control systems.

Aeration fans and aeration rates can be controlled through on/off cycling or modulating control. The aeration control system can vary in complexity from manual temperature probes (see Figure 7.1) and manually switching the fans on/off to on-line temperature monitoring with a supervisory control system.

In the manual temperature control mode, temperature monitoring is accomplished using temperature probes, inserted in at least three points in the pile, followed by manual switching of the fans or timers. This method is generally used in small composting facilities where door generation and control is not a major issue.

Larger composting facilities generally utilize an automatic temperature monitoring system to control aeration rates and odor control operations (see Figure 7.2). The temperature monitoring system consists of thermocouples embedded in the pile (see Figure 7.3), which transmit temperature readings to a central microprocessor/PC and display unit (see Figure 7.4). The microprocessor regulates the fan on/off cycle time to keep tempera-

TABLE 7.1. **Aeration System Process Design Parameters.**

Parameter	Value
Biosolids Characteristics	
Design capacity, dtpw (avg. month)	122.5
Design capacity, dtpd (avg. month)	17.5
Total solids, %	25
Volatile solids, % of TS	60
Biodegradable volatile solids, % of VS	50
Oxygen demand, lb/lb BVS	2.0
Heat release rate, Btu/lb oxygen	5,866
Amendment Characteristics	
Total solids, %	55
Volatile solids, % of TS	90
Biodegradable volatile solids, % of VS	10
Oxygen demand, lb/lb BVS	1.2
Compost Mix/Pile Characteristics	
Volumetric ratio, amendment/biosolids	2.5:1
Initial total solids, %	40
Final total solids, %	55
Exhaust air temperature, °F	130
Water of reaction, lb water/lb BVS	0.6
Minimum cycle time, days	21
Total no. of cells	19
Idle space, ft	40
No. of active cells	17
Curing Pile Characteristics	
Initial toal solids, %	55
Final total solids, %	55–60
Exhaust air temperature, °F	100
Cycle time, days	30
Total no. of cells	21
Idle space, ft.	40
No. of active cells	19
Ambient Conditions	
Design temperature, °F	95
Relative humidity, %*	50

*Based on American Society of Heating, Refrigerating, and Air Conditioning Engineers and data for Norfolk, VA.

Source: HRSD/Black & Veatch Engineering Study (1993); used with permission.

tures within a preset range. A record chart is printed (stylus-type) continuously while system is on line (see Figure 7.5). Location and spacing of the thermocouples within the pile are determined to achieve optimum aeration for that area or state of composting. For example, the first three or four cells that have the greatest microbial/composting activity require aeration only if the temperatures exceed or fall below the range for optimal composting.

Figure 7.1 A simple manual temperature probe embedded in compost pile.

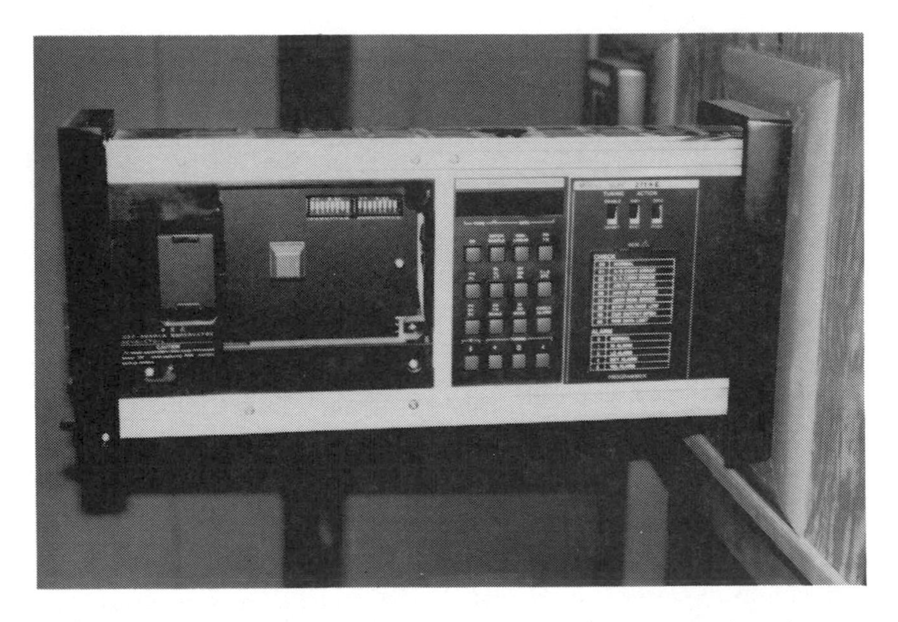

Figure 7.2 Part of the automatic temperature monitoring system used to control aeration rates.

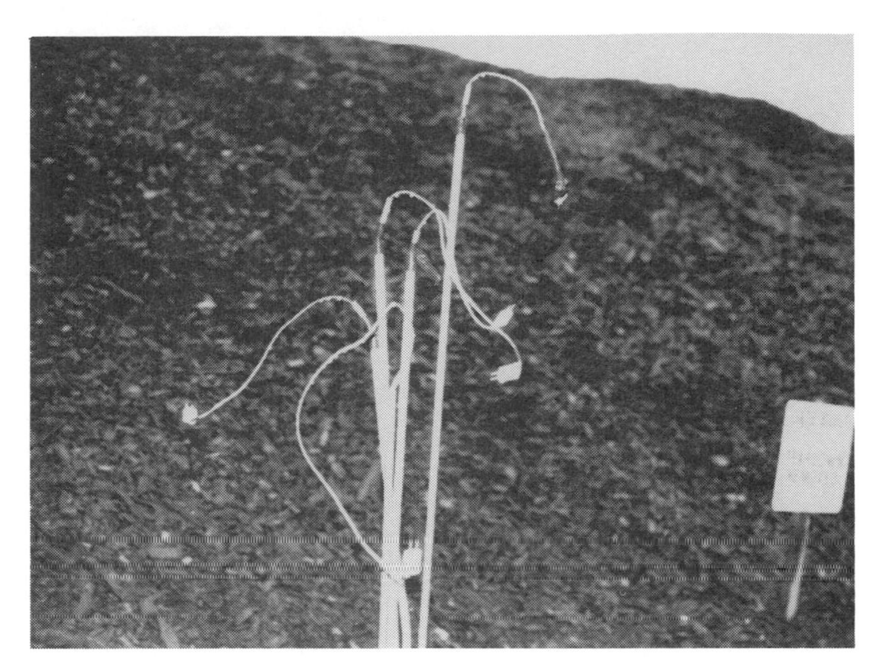

Figure 7.3 Thermocouple connection wiring used in temperature monitoring system.

Figure 7.4 Central microprocessor display unit with recording tape.

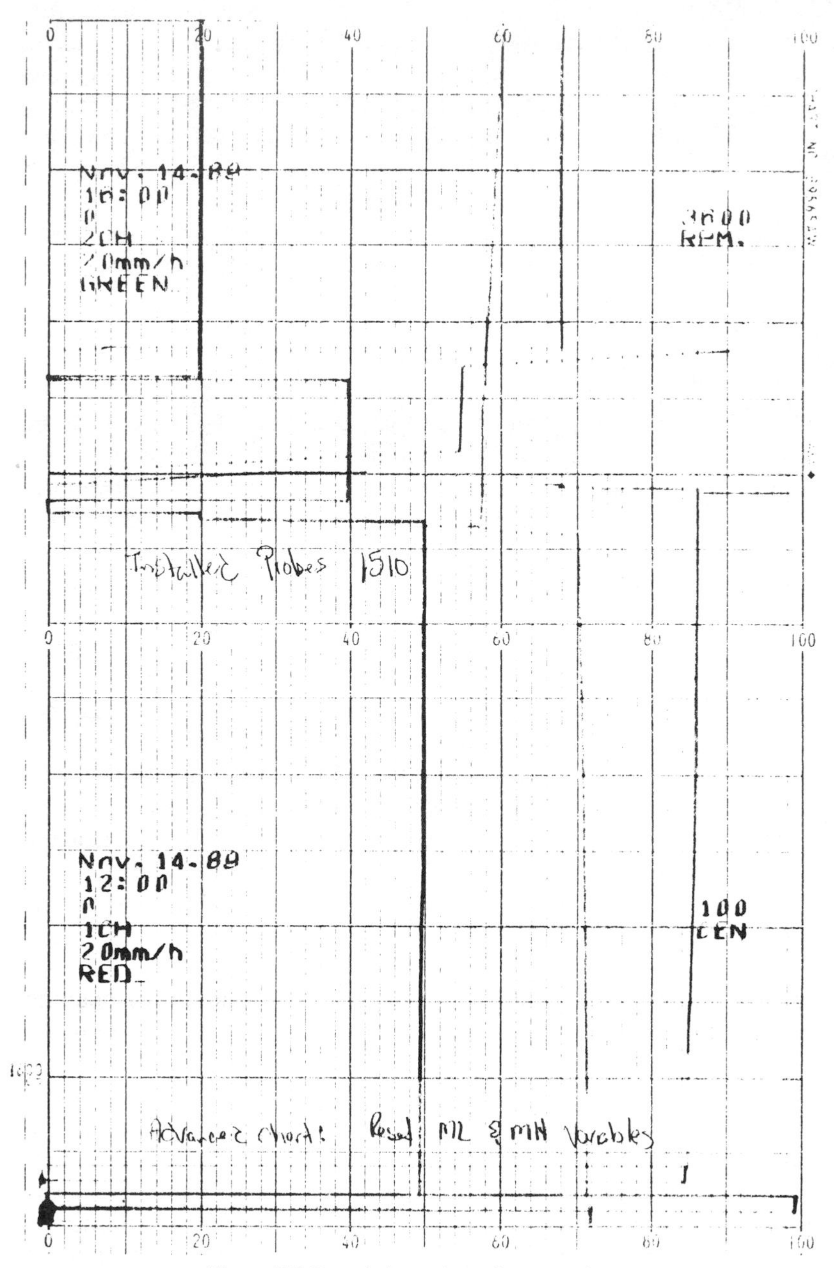

Figure 7.5 Record chart of aeration control process.

Because of its size and complexity, the ASP model composting facility uses an automated temperature monitoring and aeration control system. The time, labor, and instrumentation used to monitor each pile daily, and the accuracy of the temperature records, preclude manual monitoring of the piles. Monitoring requirements for this facility include air temperature withdrawn from composting cells, pressure in the air piping header, and air flow to the odor control system.

When designing a biosolids composting facility, several fan control system options are available. In the following, two options are discussed: modulating and on/off control.

Modulating Control

For the individual fan option, adjustable-frequency drive fans would be utilized, for the central fan option, modulating valves for each aeration cell are utilized. The individual fan or valve is controlled from temperature readings in the suction piping header. The temperature adjusts the variable-speed drive or valve position to maintain a setpoint temperature parameter. A supervisory control system adjusts temperature setpoints to account for cell age and balance of air flow to the odor control system.

There are several advantages to modulating control: (1) large temperature swings within the cell are eliminated; (2) the aeration rate closely matches process requirements; (3) fewer odors may be generated; and (4) odor surges are less likely to overload the odor control system. The disadvantages of modulating controls include (1) increased capital costs; (2) higher maintenance; and (3) higher skilled labor for control system debugging and programming changes. Modulating control is necessary for the curing pile because the airflow rate is not temperature dependent.

On/Off Control

The on/off control option involves cycling a fan on and off, or opening or closing a valve. The fans or the valves are cycled using a programmable logic controller (PLC) based on time and temperature readings in the suction header, as well as the flow balancing requrements of the odor control system. The timer has two adjustments: time between fan cycling and the minimum duration the fan operates during each cycle. After the minimum time expires, the temperature control takes over. If the temperature exceeds the setpoint, the fan continues to operate. The PLC also limits the number of fans in operation to keep the airflow to the odor control system relatively constant.

The advantage of this type system is that the controls are less expensive than modulating controls. However, the disadvantages of the system include: (1) the controls appear to be random; (2) odors may build when a cell is not aerated; and (3) pile temperatures may rise to unacceptable levels. On/off controls for the curing pile can be either manual or controlled from a PLC.

ASP MODEL TEMPERATURE MONITORING SYSTEM AND CONTROL

Compost Temperature Monitoring and Control

A resistance temperature detector (RTD) element is provided in the suction piping from each of the aeration cells (i.e., one cell per fan). Temperatures of each cell are controlled by starting and stopping the constant-speed fans associated with the cell. A PLC performs the control logic and interfaces between the fans and the data acquisition system.

The PLC maintains a temperature table of the latest valid temperatures from each of the aeration cells. The PLC only updates the temperature table value if the associated fan has been on for a minimum of five minutes.

At a given frequency, for example every fifteen minutes, the PLC scans the temperature table and selects the highest pile temperatures. The fans associated with these temperatures are selected to operate. The PLC also performs a sequence check to identify the next fan in a continual rotating sequence. This scheme is intended to operate all the on-line fans for at least 15 minutes every couple of hours. For example, assuming that nineteen cells are in operation and that ten fans are on at any one time, there are nine cells that must be rotated through the "next off-line fan sequence." For a 10-minute rotation period, all off-line fans would be operated at least once every two hours (nine off-line fans × 10 minutes' rotation). The overall result would be that every 10 minutes another off-line fan would be started and the on-line fan for the coolest cell of operating fans would be stopped. Thus, a fixed number of fans would be operating at one time.

Data Acquisition and Supervisory Control System

A personal computer-based data acquisition system is provided for monitoring and for composting operation. The personal computer (PC), located in the administration building, displays and records temperatures as well as the status and alarms for the process equipment. The PC allows operator interface and the ability to access and change temperature set-points as well as control logic modifications.

SUMMARY OF AERATION DESIGN CONSIDERATIONS

An aeration system for composting should not be installed without careful planning. An unplanned, random arrangement of piping and assorted fans guarantees results that are less than hoped for. Aeration system design has as its primary intent to provide a simple (but workable) and cost-effective method of aerating the composting and curing piles while incorporating maximum flexibility in the process controls. The recommended aeration system design protocol should provide for the following:

- Concrete aeration trenches with peforated aeration piping. Trench bottoms should be sloped to drain to a leachate/washdown collection sump. The aeration trenches should be spaced at about 8-foot intervals.
- Individual fans should be provided for each pair of aeration trenches. A header will connect the two aeration trenches to each fan. Separate shut-off valves will be provided for each trench. Each fan should be at least 2,400 cfm (i.e., if the ASP model is the desired design type).
- A central fan system should be provided for the curing pile.
- A temperature feedback and aeration rate control system, including a microprocessor to control the fans and a supervisory control system to optimize the flow of air to the odor control system, will be provided.

DESCRIPTION OF ASP MODEL AERATION SYSTEM FOR COMPOSTING

As stated, the ASP model composting facility uses the static aerated pile method of composting. In general, the homogenized mixture of bulking agent (coarse hardwood woodchips) and dewatered biosolids is piled by front-end loaders onto a large concrete composting pad where it is mechanically aerated via PVC plastic pipe embedded within the concrete slab. This ventilation procedure is part of the 26-day period of “active” composting, when adequate air and oxygen are necessary to support aerobic biological activity in the compost mass and to reduce the heat and moisture content of the compost mixture. A compost pile without a properly sized air distribution system can lead to the onset of anaerobic conditions and the appearance of putrefactive odors.

A diagrammatic layout of the ASP model composting pad is shown in Figure 7.6.

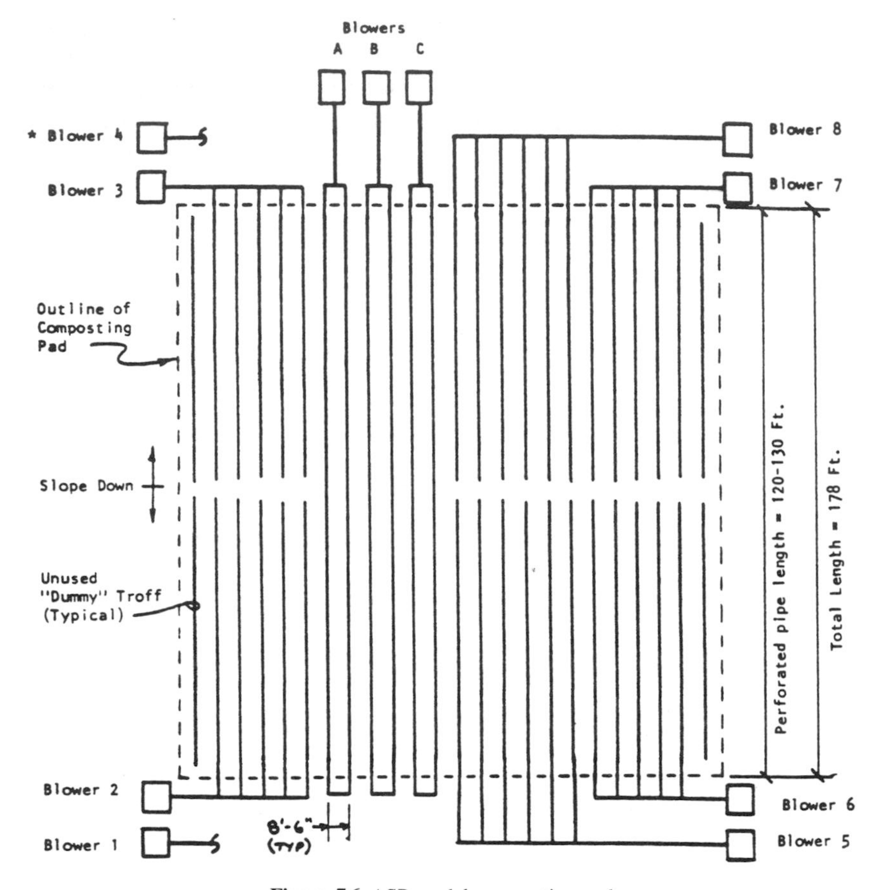

Figure 7.6 ASP model composting pad.

The overall composting pad area is approximately 200 feet by 240 feet consisting of eleven blowers and twenty-four pipe troughs (troffs). Blowers 1–8 and associated in-slab distribution piping are part of the original positive aeration system installed in 1980. When it was determined that the blowers were undersized for future growth and for minimizing odor production, the system was modified for an engineering pilot study.

The modifications included installation of blowers A, B, and C. These three blowers are 20 hp 2,400 cfm, variable-speed drive units capable of operating in either the positive or negative aeration mode. Blowers A, B, and C are each connected to two piping troughs that run the full length of the pad as shown in Figure 7.6. The two troughs are connected at the opposite end of the composting pad to create an "aeration pipe loop." The troughs are also part of the original 1980 design.

The other original eight blowers are rated at 3 hp 1,200 cfm and are arranged one blower per six troughs at half length, feeding 200 cfm per trough (see Figure 7.7). These blowers can be operated in the positive or negative aeration mode. Aeration piping within the six pipe troughs is perforated PVC plastic pipe, 6 inches inside diameter and 1/4 inch wall thickness (see Figure 7.8). Perforation holes/orifices vary in size from 7/32 inch to 1/2 inch, increasing in diameter as the distance from the blower increases. Figure 7.9 shows a schematic representation of the perforated hole pattern and the aforementioned "aeration pipe loop." Figures 7.10 and 7.11 show a series of photographs of the actual installation of perforated PVC pipe used for aeration in a typical aerated static pile type composting facility.

The variable-speed motor drives installed with blowers A, B, and C are controlled by five thermal probes mounted at various depths in the compost pile and various parameters are fed back to the recorder whereas the other eight blowers are constant-speed, controlled by a timer that cycles them on and off. To ensure optimum composting operations, it is important to verify that these thermal probes are calibrated on a regular basis (see Figure 7.12). In the constant-speed system, thermal probes are installed but all readings are taken and recorded manually (see Figure 7.13).

(a)

Figure 7.7 (a) One of eight 1,200 cfm blowers used by ASP model. (b) Closeup view of 1,200 cfm blower.

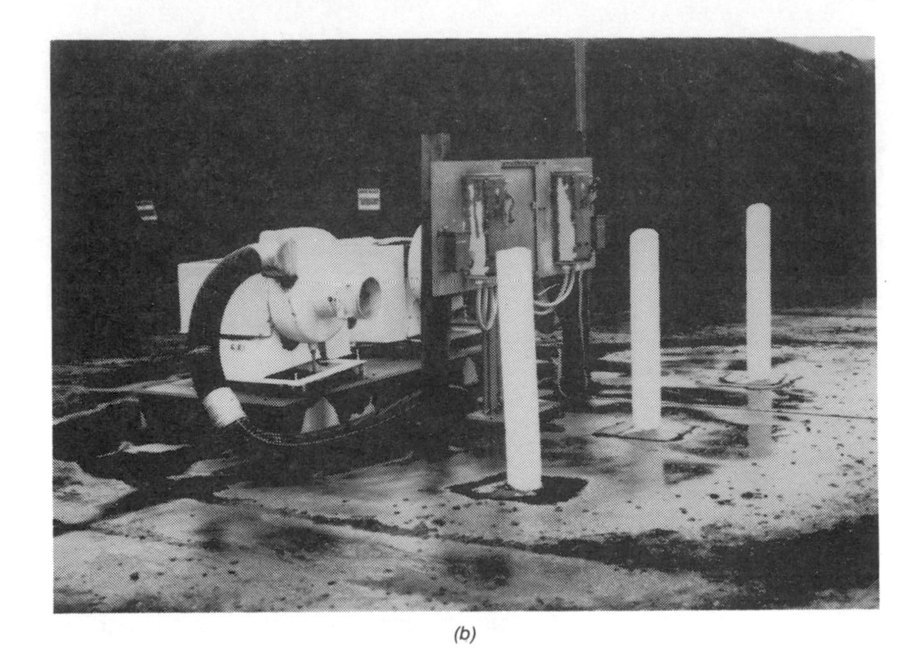

(b)

Figure 7.7 (continued) (a) One of eight 1,200 cfm blowers used by ASP model. (b) Closeup view of 1,200 cfm blower.

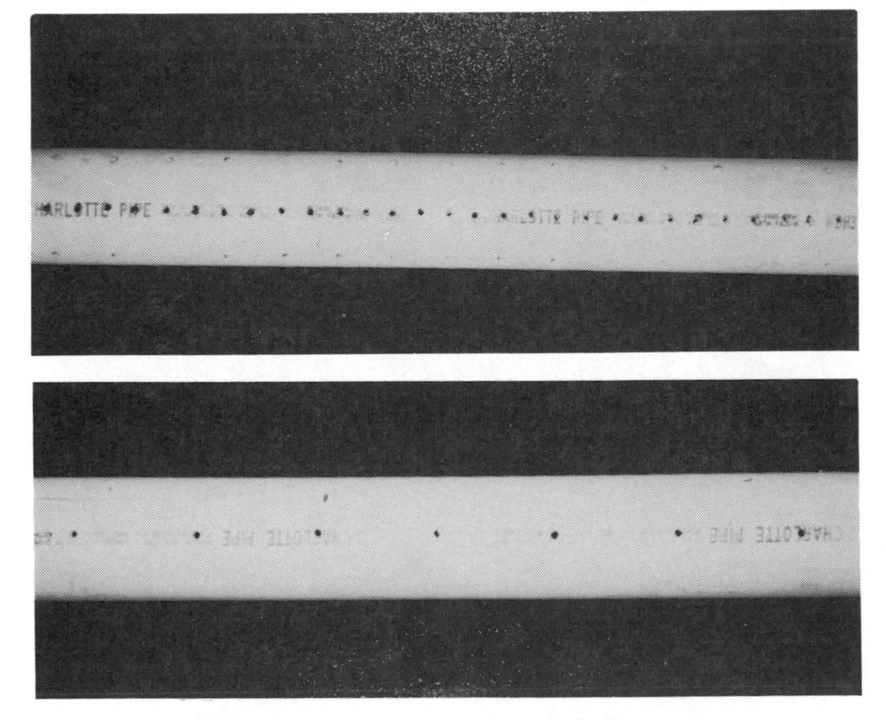

Figure 7.8 PVC pipe with two different perforation patterns.

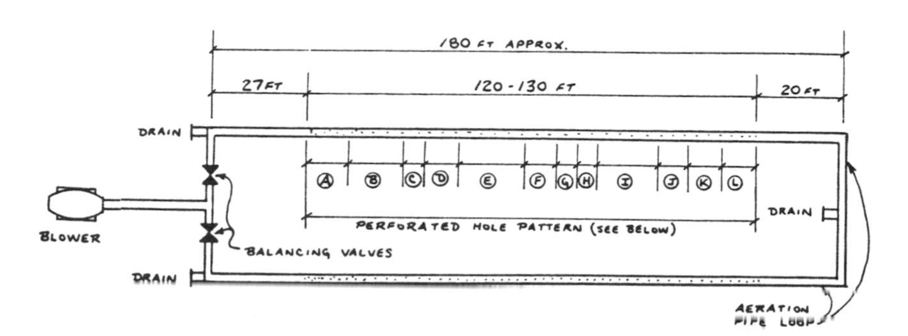

Section	Length of Section (Feet)	Hole Size (Inches)	Hole Spacing (Holes/Ft)
A	10	7/32	5
	10	5/16	2
B	15	1/4	6
	15	5/16	2
C	5	1/4	8
D	10	9/32	7
E	20	5/16	7
F	10	11/32	7
G	5	11/32	8
H	5	11/32	6
I	20	3/8	8
J	10	7/16	8
K	10	1/2	12
L	10	1/2	18

Figure 7.9 Perforated aeration PVC pipe hole pattern used in a typical ASP model composting facility.

Figure 7.10 Installation of perforated PVC aeration pipe used in a typical ASP model composting facility.

Figure 7.11 PVC aeration pipe in troughs and connection to 20 hp 2,400 cfm blowers.

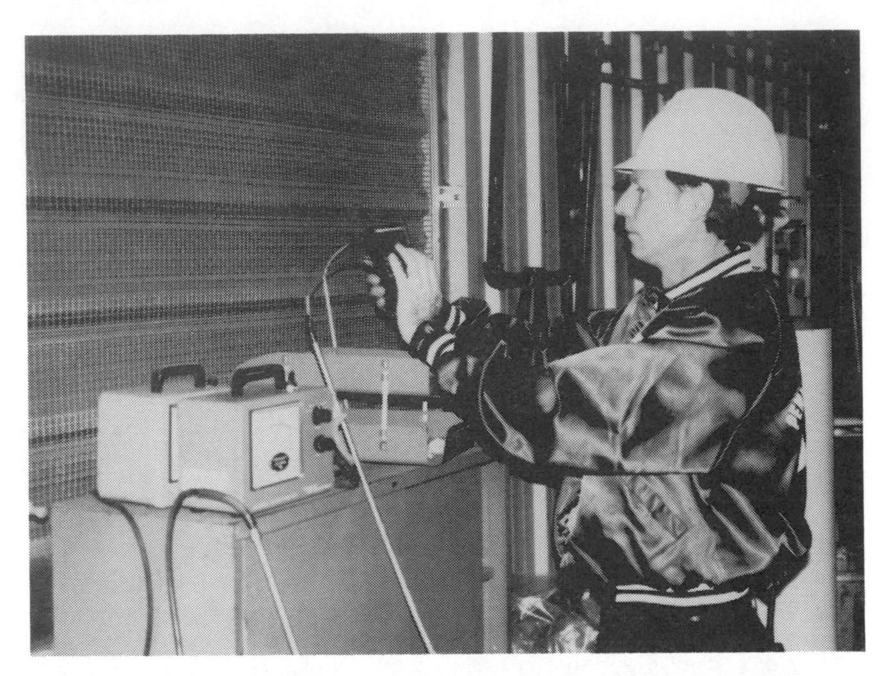

Figure 7.12 Compost technician verifying calibration of temperature probes.

Figure 7.13 Compost technician taking temperature readings.

For water and leachate drainage purposes, all aeration piping within the troughs slopes downward, with the highest point at the center of the composting pad. Drain caps at each end of the pipe length are manually removed on a regular basis so that any build-up of debris or moisture does not interfere with the airflow.

The construction process involved in building the compost pile will be covered in detail later in this chapter, but for now a few key points should be made. For example, prior to the piling of the mixture on the composting pad, an 18-inch layer of woodchips is used as a base material. The primary purpose of the woodchips base is to keep the composting mixture clear of the aeration pipes, which reduces clogging of the air distribution openings in the pipes and allows free air circulation. A secondary benefit is that the woodchips insulate the composting mixture from the pad. The compost pad is like a heat sink, and this insulating barrier improves the uniformity of heat distribution within the composting mixture.

A cross-sectional view of a typical composting pile and the embedded aeration piping used at the ASP model composting facility is shown in Figure 7.14. A section sketch of the installed 6-inch PVC piping within the trough and trough dimensions is shown in Figure 7.15. Figure 7.16 shows typical connection points from the main loop going into various sections of each pile.

CONSTRUCTION OF COMPOSTING PILE

Aerated static piles may be constructed in individual or extended piles. (The ASP model used extended piles.) The following discussion details the procedure used in their formation.

COMPOST PILE FORMATION PROCEDURE

(1) Check to ensure the nonperforated section of the aeration pipe extends at least 8–10 feet under the slope at each end of the compost pile. This practice is necessary to prevent short-circuiting of air, which could result in "cold spots" with inadequate pathogen destruction.

(2) Check to ensure aeration pipe is not damaged. Replace pipe if necessary.

(3) Fill the trough area around the aeration pile with a suitable bulking agent such as woodchips. Replace when bulking agent or composted material becomes compacted in troughs.

(4) Place a 3- to 18-inch base layer of bulking agent in an area approximately 4 feet on either side of the aeration troughs. The purpose of

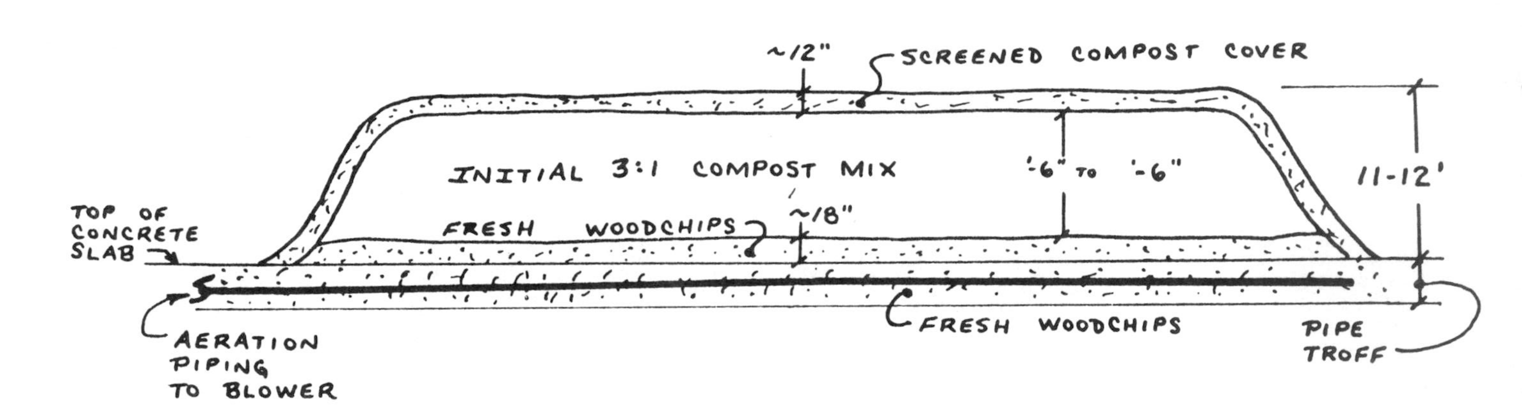

Figure 7.14 Cross-section of compost pile. Source: HRSD/Black & Veatch Engineering Study (1993); used with permission.

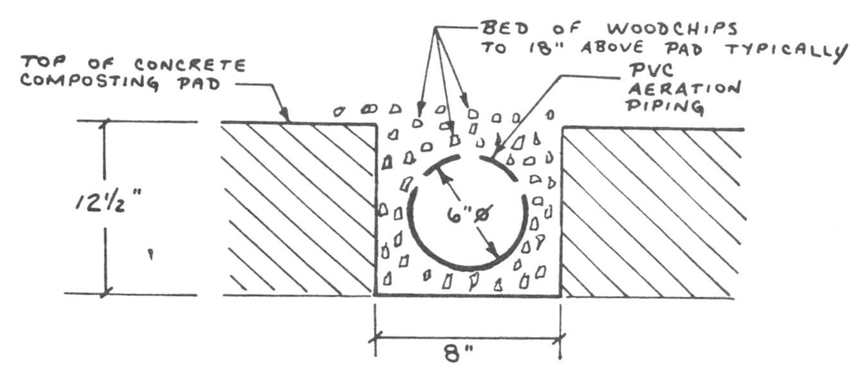

Figure 7.15 Detail of aeration piping in trough. Source: HRSD/Black & Veatch Engineering Study (1993); used with permission.

installing a bulking agent base is to improve air distribution, absorb moisture, and prevent clogging of the trenches.

(5) Construct over the bulking agent base with a front-end loader an initial pile with triangular cross-section to a convenient height of 7–10 feet. Be careful not to compact the compost mix. A compacted mix will not compost properly. If compost pile is constructed out-

(a)

Figure 7.16 (a) Trough with aeration pipe and lateral connection to various sectors of compost pile. (b) Aeration trough with covers in place.

(b)

Figure 7.16 (continued) (a) Trough with aeration pipe and lateral connection to various sectors of compost pile. (b) Aeration trough with covers in place.

doors, care should be taken while forming piles to ensure that tops of piles form peaks or rounded dome-like tops to allow for rainwater runoff; otherwise, unwanted moisture may pool in pile tops, retarding the composting process.

(6) Blanket both ends and the exterior side of the first pile with either 8–12 inches of cured screened compost or 16–20 inches of unscreened compost or bulking agent. The purpose of this "blanket" is to provide insulation and prevent the escape of odorous gases.
(7) Form subsequent piles parallel to the first. Extend the base following procedures 1–4 outlined previously. Then loosely place compost mixture next to the previous pile to form an extended pile with a trapezoidal cross-section. As the piles are made, blanket the tops and ends with 8–12 inches of cured screened compost or 16–20 inches of unscreened compost or bulking agent. Hand rake the top and sides of the piles smooth to prevent water pockets from forming. Sweep excess woodchips from around the base of the piles.
(8) At the end of each day, dust the uncovered side of the compost pile with approximately 1–3 inches of screened or unscreened compost or bulking agent for overnight control.
(9) Place appropriate sign on pile to indicate date started and date to be taken down (26 days).
(10) Close aeration dampers in troughs that have no compost piles.

A cross-section of an extended pile is shown in Figure 7.17 and a schematic of the fully constructed ASP model compost extended pile is shown in Figure 7.18.

COMPOST PILE OPERATION PROCEDURE

(1) Set blowers to operate intermittently. Aeration rates may vary from 200–1,200 cf/hr/dry ton. Generally, the blowers operate 3–25 minutes each 1/2 hour cycle. The blower operating cycle should be adjusted depending on interior oxygen levels. As a guideline: If the oxygen level is $<5\%$, blower on time should be increased; if oxygen level is $>15\%$, blower on time should be decreased.
(2) Check all blowers each day to ensure they are operating correctly.
(3) Adjust aeration dampers as necessary to ensure even distribution of air.
(4) Check all drains each day to ensure they are operating properly. Drain water from 8-inch fiberglass headers as necessary.
(5) Compost is to be sampled and tested as directed. *Note:* Sampling and testing procedures will be discussed in detail later.

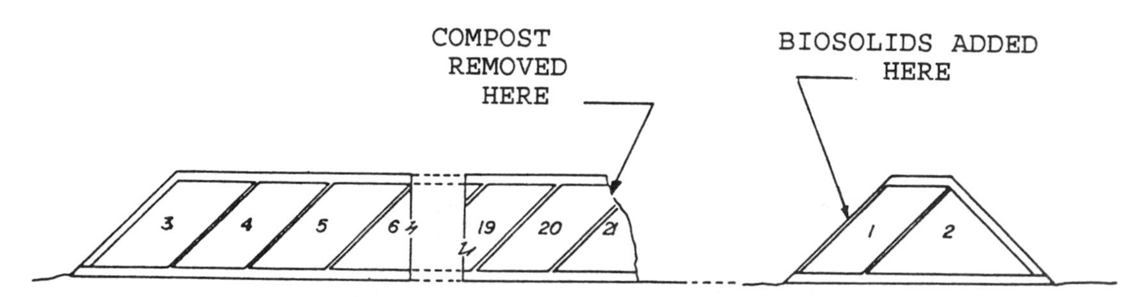

Figure 7.17 Cross-section of an extended pile with typical sequence of biosolids additions to the pile. Numbers indicate the age of the compost in days.

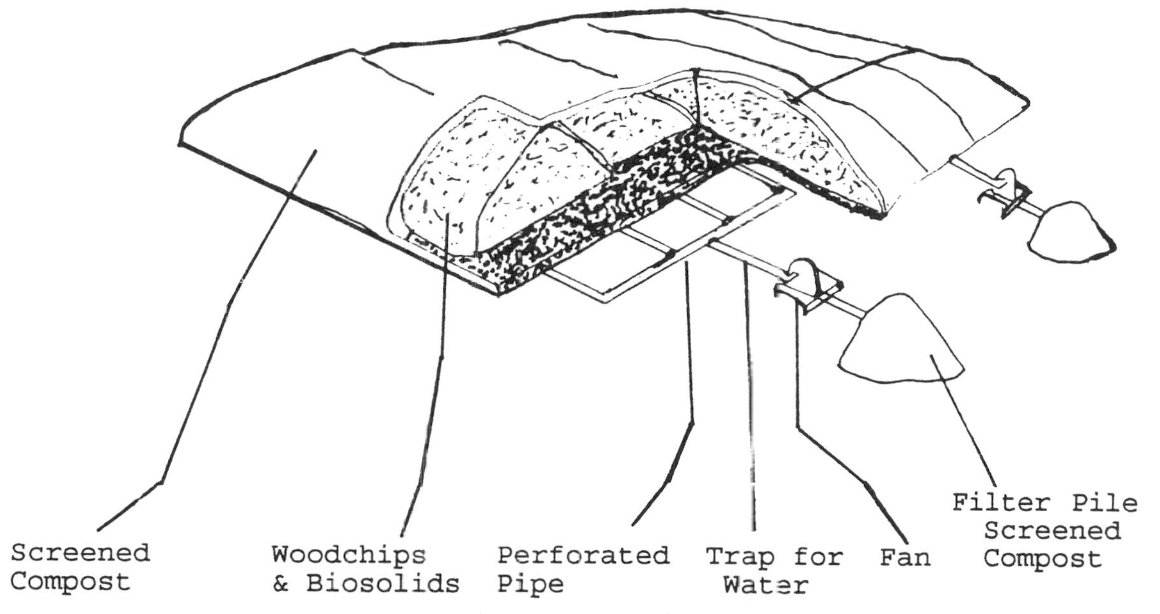

Figure 7.18 Schematic diagram of extended aerated static pile showing construction of pile and arrangement of aeration piping.

(6) Temperatures should increase to 50 °C within a few days.

(7) If temperatures of 55 °C have been maintained at all pile monitoring points for a minimum of 3 days, the compost pile can be removed after 21–26 days. *Note:* Piles are to be removed with front-end loader. Care should be taken to break into pile on upwind side.

(8) Compost that has not maintained pile temperatures of 55 °C for at least 3 days, must be recycled back through the composting process.

CHAPTER 8

Curing and Drying

AFTER 26 days of composting, the next step in the process is either drying or curing, or a combination of both. In some facilities drying is an optional stage, but is usually necessary if the compost is to be recycled as a bulking agent or if screening is required.

The ASP model composting facility combines the curing and drying stages into one stage. As stated earlier, the curing/drying stage occurs after twenty-six days of aerated composting. After the curing/drying stage is complete, the curing/drying pile is dismantled by front-end loaders, picked up by front-end loaders, and deposited into the feed hopper for the screening device. In some cases, the compost is screened first and then cured/dried. However, the ASP model screens after the curing/drying stage. As was the case with building the composting piles, front-end loaders are used to construct the curing and drying piles. Unlike the composting pile, the curing/drying pile does not have an insulation cover. After thirty days of aerated curing/drying, front-end loaders dismantle the drying/curing pile from its trailing edge.

For the purpose of illustration, the curing and drying processes are described separately.

THEORY OF OPERATION

CURING

During the curing stage, compost is stabilized as microorganisms metabolize the remaining nutrients in the biosolids-compost mixture. Curing is required for a minimum of thirty days prior to final sampling/testing (to be discussed later) and distribution. Curing and any subsequent storage is generally considered to be an extension of the aeration process and is associated with elevated temperatures, although at

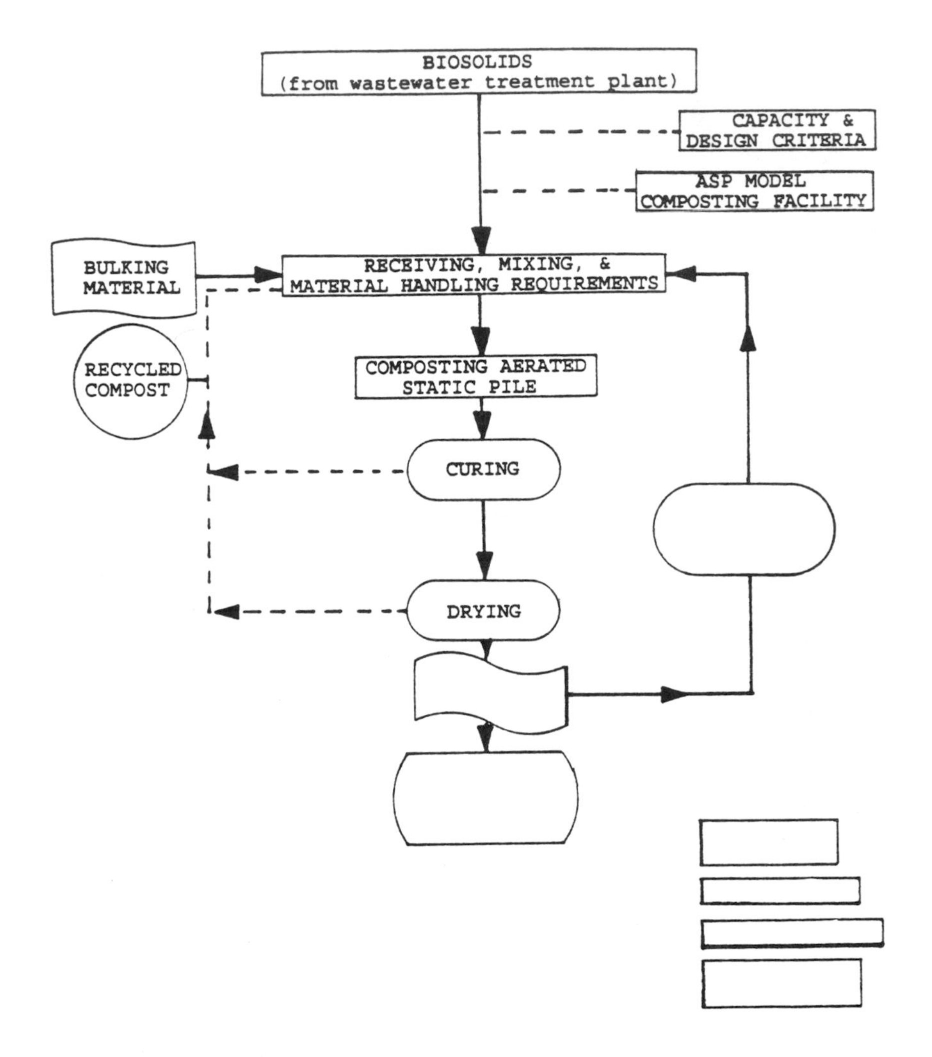

somewhat lower than the average temperatures attained during the initial composting. Curing ensures total dissipation of odor-causing gases and allows for destruction of any remaining pathogens. Once the curing stage is completed, the finished compost product should not have an unpleasant odor.

DRYING

As stated, drying is optional, but is usually necessary if compost is to be recycled as a bulking agent or if screening is required. Normally, drying occurs in a designated shed or structure where the roof will protect the compost from inclement weather. However, if weather conditions permit, drying can occur on any hard-surface area of the facility.

Drying is accomplished by drawing or blowing air through the aeration pipes located in the troughs, or by mechanical mixing with a front-end loader, disc, or composting machine. A combination of mechanical mixing and aeration can also be used.

PROCESS OPERATION

CURING OPERATION

Front-end loaders or trucks can be used to transport material from the compost pad, drying, or screening area to the curing area. The curing area is generally smaller than the composting area. The compost must be stored for thirty days prior to distribution. When constructing the curing piles, it is important (when curing occurs outdoors) to shape the tops of the curing piles so that they are domed or rounded in order to prevent water buildup.

DRYING OPERATION

Drying can occur under cover in the drying shed or on any concrete surface in the open.

Drying under Cover (Procedure)

(1) Check for damaged aeration pipe and replace if necessary.
(2) Fill the trough area around the aeration pipe with a suitable bulking agent (e.g., woodchips). Replace when bulking agent or composted material becomes compacted in troughs.
(3) Replace composted material over troughs in pile of triangular cross section.
(4) Open aeration damper.
(5) Check all blowers and drains each day to ensure they are operating correctly.
(6) Drain water from header as necessary.
(7) Blowers can be adjusted to force air out or draw air through pile.
(8) Drying can be aided by periodic turning with the composting machine or front-end loader.
(9) Drying is complete when compost has a moisture content suitable for screening.

Drying in Open Air

(1) Observe weather conditions when drying outside. Outside drying should not be done if precipitation is anticipated.
(2) Spread pile to facilitate drying. Drying can be aided by periodic turning with the composting machine, front-end loader, or disc.
(3) Drying is complete when compost has a moisture content suitable for screening.

CHAPTER 9

Screening

THE final step in the composting process is the screening step. Specialized screening equipment is required to separate the compost product from the woodchips after curing and drying. The recovered woodchips are recycled, while the screened compost product is moved to the distribution area where it is eventually marketed in bulk or in bags.

SCREENING

In order to properly screen the cured and dried compost for woodchip recovery and production and marketable compost product, it is important to have a feed compost that is 55–60% TS. Some screening devices can function with a feed compost solids concentration as low as 50% TS, but the screening mechanism can clog rapidly, causing the unit to be stopped frequently for cleaning (Finley, 1996).

There are several types of compost screening devices: stationary screens, vibrating screens, shaker screens, and trommel screens. For the purposes of this text, shaker- and trommel-type screens will be discussed.

SHAKER SCREENS

As their name indicates, shaker screens move with an up-and-down shaking motion. This action sifts the compost and bulking agent material in such a way that the compost is allowed to pass through the screen whereas the woodchips are separated and placed in another location. Screen size is dependent upon the size of the bulking agent.

The major disadvantages of shaker screens is their tendency to clog and the extensive amount of maintenance required (as with many other types of mechanized machinery) to keep them on line. In order to reduce clogging problems, it is important to maintain a clean screen deck. If screens

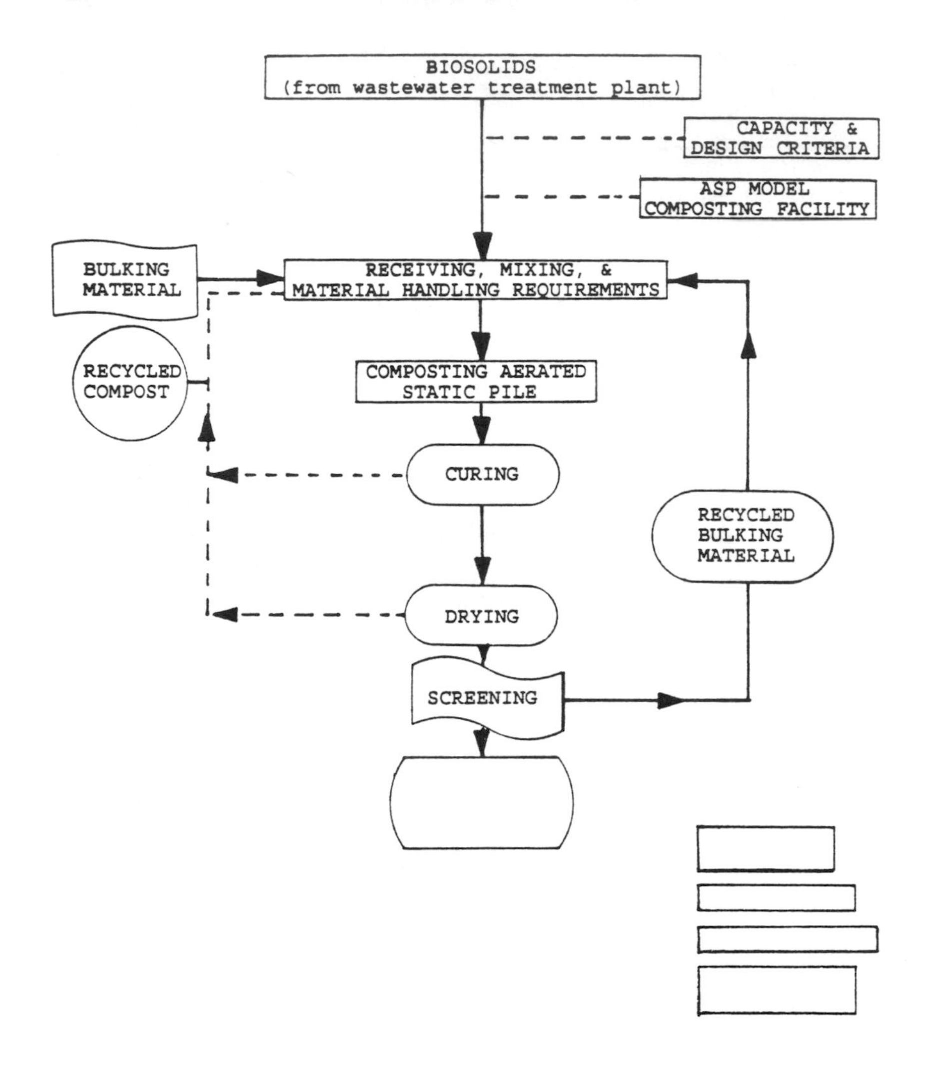

are allowed to clog, fine material will not pass through but simply be recycled into the composting process with the woodchips. This is neither desirable nor cost effective, since extra material handling will result.

TROMMEL SCREENS

Trommel screens are long, cylindrical rotating screens that are usually placed on an angle so that materials flow through them (see Figure 9.1). The materials rotate with the trommel. Within the rotating trommel drum, the compost product and bulking agent are separated by a tumbling action, similar to that of a clothes dryer. Compost materials smaller than the grate size fall through, while bulking agent is retained in the screen until discharged at one end. Screened compost from the trommel is usually collected on an underbelt conveyor system that travels the length of the trommel drums and falls through a chute onto a collection conveyor. Conveyors transport both woodchips and compost from the screens. The trommel drum rotates on wheels and generally can be tilted from 3 to 12 degrees. Large brushes mounted on top of the trommel drum extend through the screen to prevent material from clogging the screen, thus making the trommel unit "self-cleaning."

When trommel screens are used, compost facilities usually install at least two units for system redundance (because of their mechanization, downtime for routine repair and preventive maintenance is necessary). Generally, front-end loaders are used to deliver cured compost to designated receiving bins. From the receiving bin, the compost is transported via a discharge screw conveyor to the feed conveyor. It is normal operating procedure to install both the compost feed and woodchip discharge belt conveyors so that they are inclined to the points required for correct process operation. Recovered woodchips are conveyed to a temporary stockpile. Front-end loaders are used to move the woodchips to receiving bins or to the covered woodchip area. The screened product collection conveyor is a level belt conveyor, which collects the screened product and conveys it to an inclined belt conveyor. The inclined conveyor transports screened compost from the compost storage area building.

SCREENING: ASP MODEL COMPOSTING FACILITY

The layout of the ASP model composting facility screening process is shown in Figure 9.2. Two rotary trommel screens (SCR-1 and SCR-2) are provided and space is available for a third screen for future capacity. The

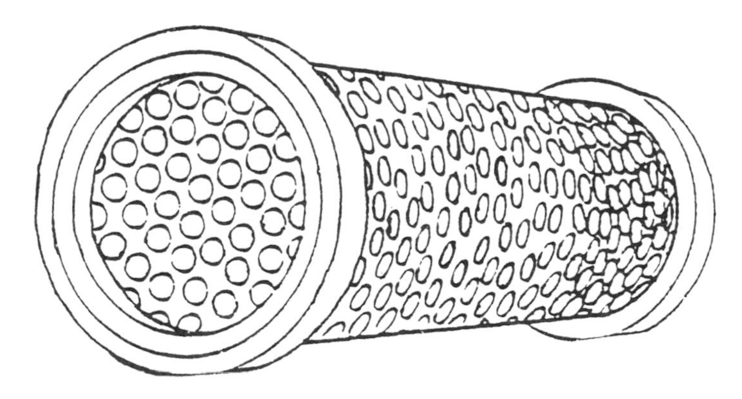

Figure 9.1 End view of trommel screen.

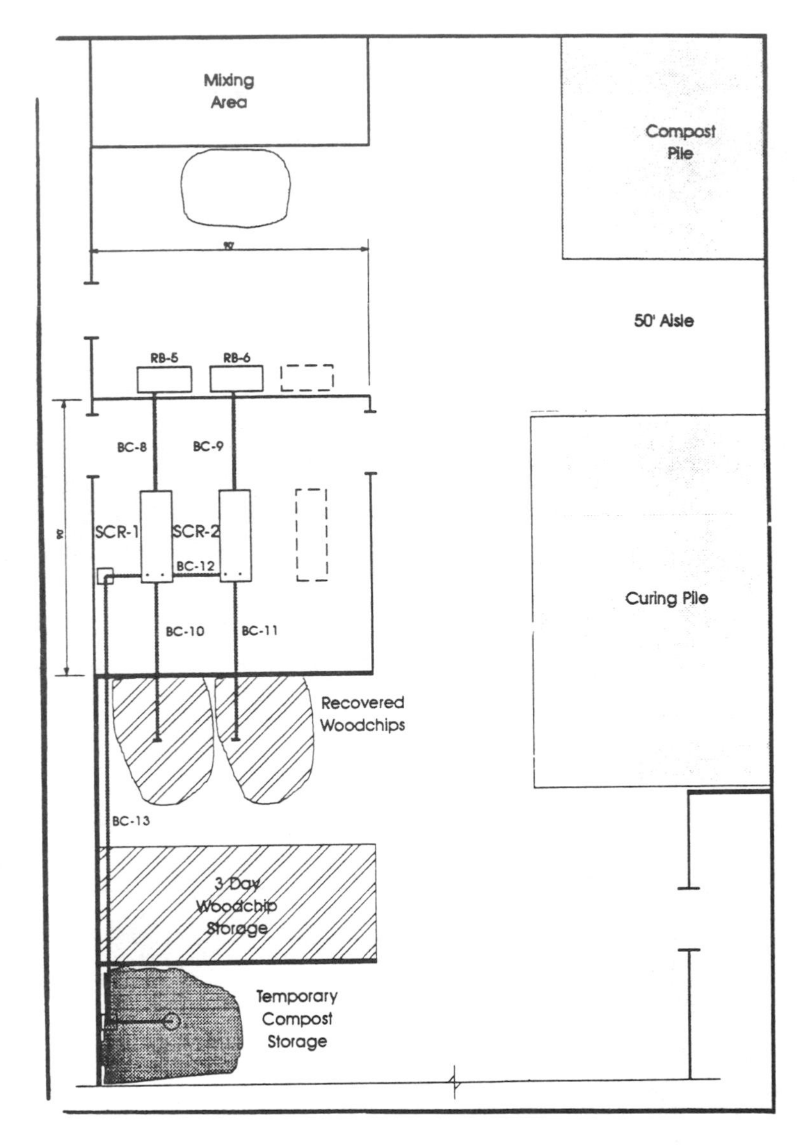

Figure 9.2 Layout of screening process at ASP model composting facility. Source: Adaptation from HRSD/Black & Veatch Pilot Study (1993); used with permission.

site layout diagram also shows the receiving bins (RB-5 and RB-6) and material conveyors (BC-8, 9, 10, 11, 12, and 13).

The screens used at this facility are Mark III power shaker compost screening devices as shown in the series of photographs in Figure 9.3(a) and (b). Screening can be conducted outside or under the cover provided by the shed.

(a)

(b)

Figure 9.3 (a) Trommel shaker screen, (b) closeup view of same screen, (c) screened compost leaving trommel screen.

(c)

Figure 9.3 (continued) (a) Trommel shaker screen, (b) closeup view of same screen, (c) screened compost leaving trommel screen.

As stated previously, the compost feed (cured/dried compost) is stored out of the weather at all times; thus, by the time the feedstock is ready for screening, the compost has had even more time to dry.

Two portable shaker screens that can be transported anywhere on the facility are available at the ASP facility. This redundancy allows for greater flexibility in operations and reduces materials handling time. The screens have two screen decks; the second screen is adjustable to any product size. This is a key feature since different compost customers require different grades of screened compost. Screening rates vary from 60 cy/hr to 120 cy/hr depending on feed compost dryness and screen deck size for each screen. Approximately 2.9 cubic yards sof screened compost is produced for every dry ton of biosolids received.

CHAPTER 10

Distribution and Marketing

The amount of compost currently produced is small in comparison to the potential demand in the U.S. Current use of compost is estimated to be fewer than 27 million metric tons per year, but studies by Buhr et al. (1993) and Slivka et al. (1992) indicate a total potential compost demand of 450 million metric tons (500 million tons) per year. The largest markets for potential increased use of compost include agriculture, silviculture, and sod production. (WEF, 1995, p. 160)

INTRODUCTION

THE ultimate goal of the biosolids-derived composting process is to produce a marketable product. Before the final compost product can be legally marketed, it must meet several regulatory requirements under federal and state guidelines. The requirements listed here are derived from the EPA's 503 regulation. State requirements differ from state to state and should be referred to for further guidance. The EPA's requirements are listed in the following:

- The finished compost product must be composted for at least twenty-one days.
- Temperature must be maintained at 55°C for three days during the composting process.
- Compost must be cured for thirty days.
- Several tests must be performed on the compost during and after the process. The compost must be tested on a regular basis for nutrients, metals, pesticides, and salmonella.
- The metals tested and their maximum levels are as follows:

Boron	100 mg/kg
Cadmium	20 mg/kg

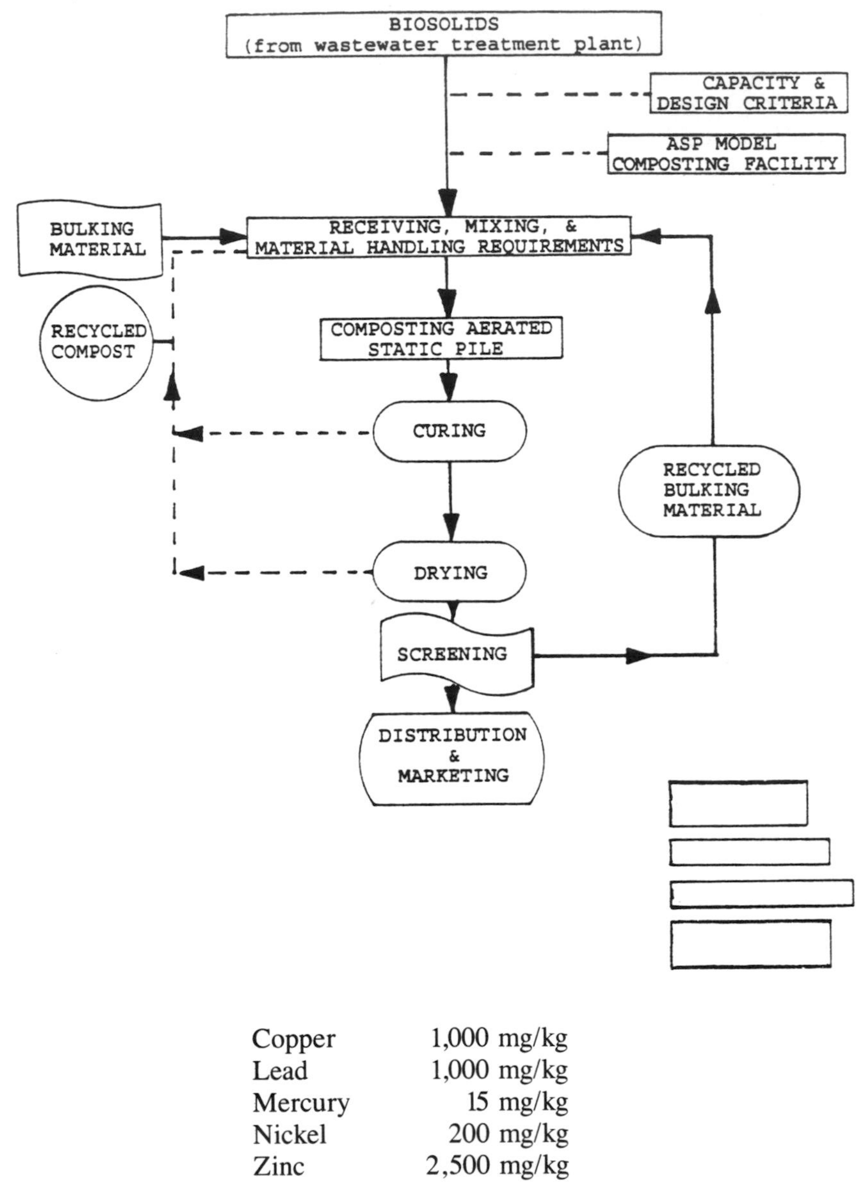

Copper	1,000 mg/kg
Lead	1,000 mg/kg
Mercury	15 mg/kg
Nickel	200 mg/kg
Zinc	2,500 mg/kg

Sampling and testing procedures will be discussed in more detail later in this text.

MARKETING COMPOST

After having produced an exceptional quality biosolids-derived compost product, the last thing one wants to do is to ask the question: What am I going to do with the finished compost (Albrecht, 1987)? The key to success for any biosolids-derived composting process is marketing (Borberg & McLemore, 1983). Fitzhugh et al. (1994) point out that in the past "the marketing of biosolids [compost] was considered an afterthought to wastewater treatment planning" (p. 72).

Fortunately, things have changed dramatically in recent years. Planners now understand that in order to ensure that the compost product can be successfully marketed, a prior-to-construction/operation study of compost marketing potential should be conducted in the local area. If the preliminary market study indicates that the potential is high for marketing the product locally, then an aggressive marketing strategy must be developed.

In order to develop a successful biosolids-derived compost marketing program, a public acceptance strategy must be developed. Kenneth Wheeler (Mendenhall, 1990) has developed a strategy for selling the sensitive biosolids marketing program. Wheeler's marketing strategy includes six steps:

(1) Prove a need—what is to be accomplished, why, and at what cost.
(2) Know the facts—find out what other facilities have done and learn from their successes and failures.
(3) Conduct research—assemble information about local attitudes.
(4) Identify friends and adversaries— find current and potential allies and adversaries. Develop a plan that optimizes relationships with allies and neutralizes opposition by opponents.
(5) Have a plan—develop clear goals.
(6) Target communication—get the right information to the right people at the right time.

After the marketing strategy is developed and put into effect, another question soon surfaces: How is a successful marketing program maintained? Albrecht (1987) answers this question by pointing out that key essentials are common to all successful biosolid-derived compost marketing programs. These key essentials are listed as follows:

- The product must have consistent quality.
- The supply of the product must be reliable.
- Increasing the variety of the finished product increases the success rate of marketing.
- The marketing institution must stand behind its product.
- There are no "pat" or set solutions for compost that are applicable to all situations.
- Marketing is a continuing effort.

Even with a sound marketing strategy and a set of key essentials to enable success, introducing compost to end users can be a difficult undertaking, especially at first (Oberst, 1996). In order to gain a better understanding of what is involved in developing a successful distribution and marketing plan for biosolids-derived compost, the following case study is presented; it details the approach undertaken by Hampton Roads Sanitation District (HRSD). According to Finley (1996), the major disadvantage of HRSD's compost marketing strategy has been that the final product cannot be produced fast enough and in sufficient enough quantities to satisfy increasing demand. That is, current demand outpaces the supply.

CASE STUDY

After a favorable preliminary marketing study, the Hampton Roads Sanitation District's composting facility was constructed and went into operation in 1981. When the HRSD facility went into operation about fifty such facilities were operating in the United States. At the present time, there are more than 200 composting facilities spread throughout the nation.

The goal at HRSD was to process approximately 15% of the total biosolids generated from nine of the district's wastewater treatment plants into a beneficial reuse compost product (Borberg & McLemore, 1983). HRSD's 12-dry ton/day composting facility has performed well; its compost product easily meets U.S. EPA's Part 503 "exceptional-quality" limits for metals and the Class A requirements for pathogen and vector attraction reduction.

HRSD makes available to the public a compost product that is sold both in bulk and bagged (40-pound bags) quantities. Rhonda Oberst (1996), HRSD's agronimist/recycling manager, points out that developing the marketing network for HRSD's compost product took a lot of legwork. Oberst worked with horticulturists at the Hampton Roads Experimental Station of Virginia Tech University (VPI) in applying HRSD's compost product in different applications to determine the best use at different quantities. Ms. Oberst understood that if she was able to compel local nurseries, horticulturists, landscapers, and the public into using HRSD's compost product, the compost would sell itself. This is exactly what occurred.

In her marketing effort, Ms. Oberst understood that to increase its marketing potential, HRSD's compost product would have to be recognized by the consumer under some catchy trade name. The trade name "Nutri-Green" was chosen and, as the saying goes: The rest is sales history.

As stated earlier, Nutri-Green is sold in two forms: bulk and bagged. Sales in bulk quantities have been overwhelming. As soon as landscapers, in particular, find out that bulk Nutri-Green is ready for sale, trucks literally line the roadway leading to the loading area. As soon as the Nutri-Green is put on the market, it is gone within less than one day (Finley, 1996).

The sale of Nutri-Green in bagged form has also been successful. The HRSD composting facility has bagging equipment installed on site. This bagging equipment includes an outside hopper with a 5-cubic yard capacity. Front-end loaders transport finished compost from the distribution pad to the hopper (shown in Figure 10.1). The bagging equipment bags the compost from the hopper at a rate of sixteen bags per minute (see Figure 10.2). The bagging equipment includes a hot-air bag sealer and a weighing scale. After bagging, individual bags are palletized and spiral wrapped in clear or black plastic (see Figure 10.3).

Under the direction of Rhonda Oberst, Nutri-Green is advertised with an emphasis on its beneficial-use applications. For instance, HRSD advertisements point out that Nutri-Green (1) increases the ability of sandy soil to hold water; (2) increases the drainage of heavy clay soils; (3) enhances aeration of heavy clay soils; and (4) provides long-lasting essential plant nutrients.

Bagged Nutri-Green is attractively packaged (see Figure 2.7). Printing on the package label emphasizes that Nutri-Green is a beneficial-reuse end product, which demonstrates progress in recycling. Moreover, advertise-

Figure 10.1 Front-end loader depositing compost product into hopper, which feeds bagging equipment for HRSD's Nutri-Green compost.

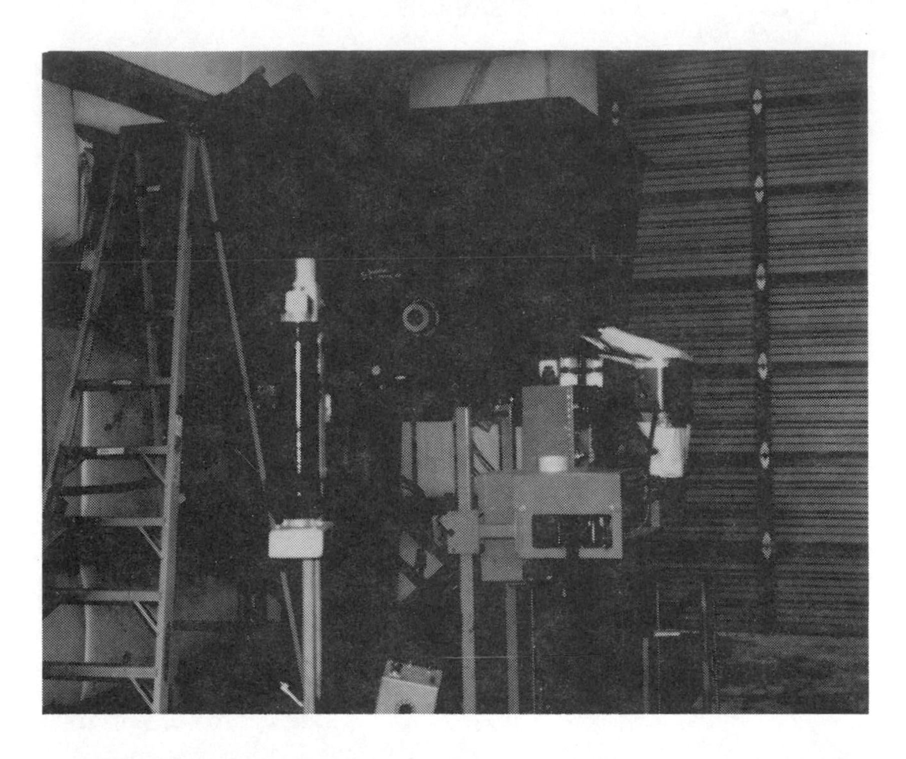

Figure 10.2 (top) Bagging equipment; (bottom) worker checking bags of Nutri-Green after bagging operation.

Figure 10.3 Nutri-Green bags that have been palletized and wrapped for shipment.

ments clearly point to the benefits of using Nutri-Green for lawns, trees, flowers, shrubbery, vegetable gardens, and house plants. The major point made is that Nutri-Green is an "All Natural Soil Conditioner" and has a guaranteed analysis of 2–2–0. More specifically, Nutri-Green's analysis is as follows:

Total nitrogen	2%
1.0% Water insoluble nitrogen	
Available potash	2%
Soluble potash	0%
pH (neutral)	7.0
Weed free	

HRSD's marketing strategy includes providing Nutri-Green customers with information on how to use Nutri-Green to its fullest advantage. For example, HRSD provides customers with the following tips: (1) how to establish a new lawn; (2) how to top-dress an existing lawn; (3) how to establish flower, vegetable gardens, and bedding plants; (4) how to mend small bare spots; (5) how to transplant ornamental trees and shrubs; and (6) how to make potting soil.

As required by the EPA's 503 rule, HRSD also provides additional information on its packages as shown in Figure 4.10.

CHAPTER 11

Sampling and Testing

INTRODUCTION

EARLIER in this text (Chapter 2), a sampling and testing procedure for determining total solids percent of concentration of processed biosolids was discussed. The goal of this testing procedure is for the solids handling operator at the wastewater treatment plant to be able to monitor the biosolids operation and to make control adjustments as needed. Such adjustments ensure production of a biosolids cake that has a total solids content in the 20 25% range (according to Baumler [1996], when centrifugation is used in dewatering, a total solids content of more than 25% is possible), which is ideal for biosolids-derived composting. This chapter will discuss other sampling and testing requirements related to processing biosolids-derived compost that is fit for use in land application.

REGULATORY REQUIREMENTS

Because of the biosolids quality requirements in EPA's Part 503 rule, sampling and testing must be conducted on biosolids-derived compost. Specifically, the EPA's Part 503 rule created standards for those pollutants on which sufficient information was available to establish protective numerical limits (Goldstein, 1993).

To be in compliance, certain parameters must be monitored. These parameters include the metals arsenic, cadmium, chromium (recently deleted as a requirement), copper, lead, mercury, molybdenum, nickel, selenium, and zinc. Moreover, indicator organisms or pathogens must be monitored. In the ASP model composting facility, pathogens are monitored; that is, *Salmonella* is monitored using most probable number or colony forming units per 4 grams of total solids. Additionally, monitoring ensures compliance with vector attraction reduction requirements to prevent the attraction of rats, flies, and other vectors.

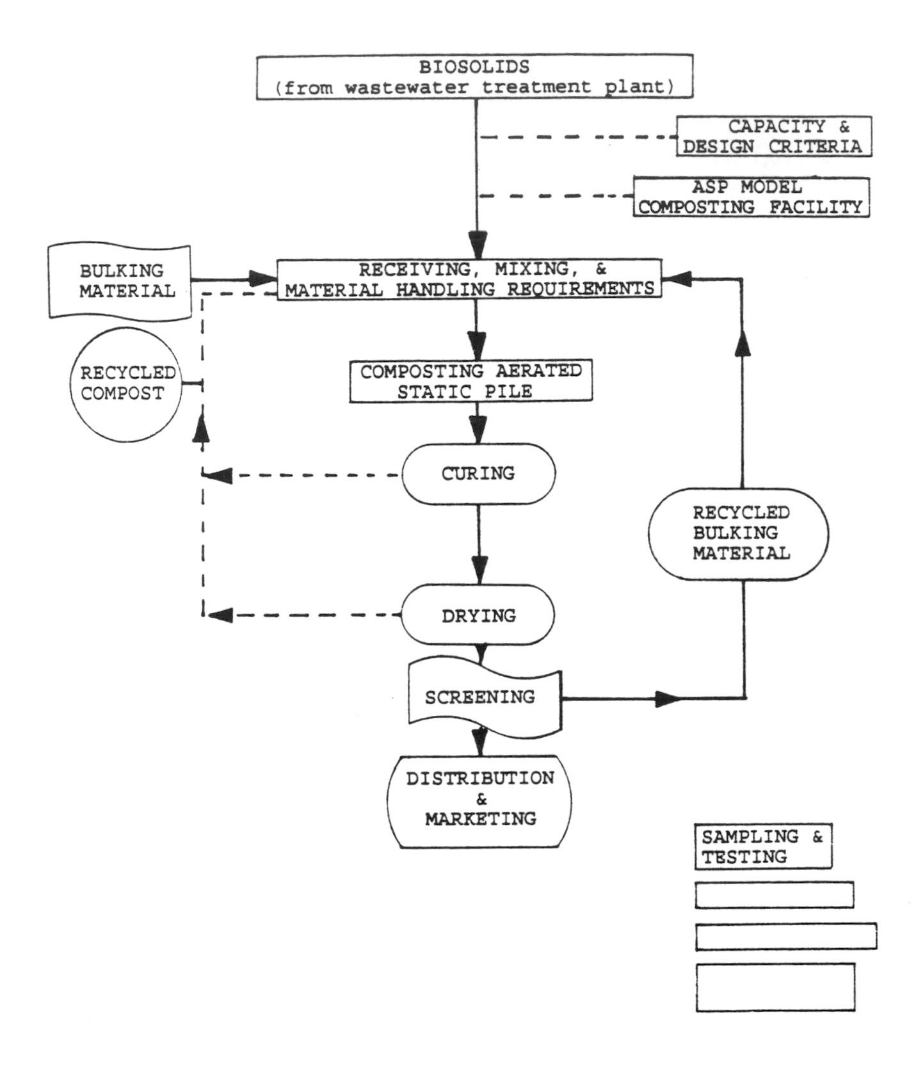

The minimum frequency of monitoring (sampling and testing) for metals, indicator organisms, and vector attraction reduction requirements is based on the amount of sewage biosolids produced annually, as shown in Table 11.1. State or other local permitting authorities not only require recordkeeping but also may require more frequent monitoring requirements than those listed in Table 11.1 (Outwater, 1994).

NUMERICAL LIMITS FOR CONTAMINANTS

40 CFR Part 503 requires the EPA to set numerical limits for contaminants in sewage biosolids to ensure that all environmental media are protected on an equal basis. Sewage biosolids sold or given away in bulk form

TABLE 11.1. Monitoring Frequency.

Sewage Biosolids Amounts (dry metric tons/year)	Monitoring Frequency
>0 to <200	Once per year
290 to <1,500	Once per quarter
1,500 to <15,000	Once per 60 days
≥15,000	Once per month

for land, lawn, and home garden application and sold or given away in bags must meet the pollutant concentration limits listed in Tables 2.5 and 11.2. If sewage biosolids meets these pollutant concentration limits, as well as Class A pathogen reduction requirements (*Salmonella*), and a vector-attraction reduction processing option, it is classified as "Exceptional Quality" biosolids and may be exempt from certain practices and requirements applicable to land-application practices.

PATHOGEN REDUCTION REQUIREMENTS

As stated earlier, EPA's Part 503 rule not only requires monitoring of sewage biosolids contamination limits as listed in Table 11.2, but also details pathogen reduction requirements. Substantial numbers of human enteric microorganisms, some of which may be pathogenic, are present in municipal sewage biosolids. As a result, the 503 rule requires that sewage biosolids receive specific levels of pathogen reduction before land application to reduce the potential for impact on human health.

TABLE 11.2. Pollutant Concentration Limits.

Pollutant	Pollutant Concentration Limits, Monthly Averages (mg/kg)
Arsenic	41
Cadmium	39
Chromium*	1,200
Copper	1,500
Lead	300
Mercury	17
Molybdenum	18
Nickel	420
Selenium*	100
Zinc	2,800

*EPA has eliminated all chromium limits for land-applied biosolids in the Part 503 biosolids rule and has increased allowable selenium concentrations in "exceptional quality" biosolids from 36 to 100 mg/kg.

Source: 60 FR 54764, October 25, 1993/U.S. EPA 40 CFR Part 503.

The Part 503 rule classifies sewage biosolids into two categories or classes, Class A and Class B. The difference between Class A and B determines the restriction on utilization. Definitions of the two classes are as follows:

(1) Class A: Biosolids meeting the requirements of this classification is equivalent to Process to Further Reduce Pathogens (PFRP) standards. The EPA allows one of six alternative methods to be used, and all require testing for indicator bacteria (e.g., fecal coliform or *Salmonella*). The indicator standard is 1,000 fecal coliforms per gram of dry solids. This testing is coupled with one of the following: pH requirements; enteric virus or helminth ova determination for low pathogen biosolids; time and temperature requirements; enteric virus or helminth ova determination for typical biosolids; PFRP; PFRP equivalent.
(2) Class B: This class is similar to Process to Significantly Reduce Pathogen (PSRP), as opposed to PFRP. The fecal coliform indicator is less than 2 million fecal coliforms per gram of dry solids. Because there is greater risk for Class B, EPA has imposed the following restrictions:
    - food crops – no harvesting after application for 14–38 months, depending on type of crop grown and how biosolids are applied
    - feed crops – no harvesting for thirty days after biosolids application
    - pasture – no animal grazing for thirty days after application
    - public access – restricted access for thirty days after biosolids application for low-exposure areas, one year for high-exposure areas
    - turf – no harvest for one year after biosolids application (Fitzhugh et al., 1994)

In this text only the Class A requirements will be addressed because they directly relate to biosolids-derived composting activities described in the ASP model composting facility process. If you use a composting method other than the ASP model, you need to review the EPA 503 rule to determine its impact on your particular biosolids-derived composting operation.

Class A Sewage Biosolids (ASP Model Composting Facility)

As stated earlier, in order to be classified as Class A, the biosolids must meet one of two critieria or one of several alternatives listed in EPA's Part 503 regulations. These criteria and alternatives are as follows:

(1) A *Salmonella* sp. density of less than 3 MPN/4 grams TS
(2) A fecal coliform density of less than 1,000 MPN/g total dry solids (TS)

Or the requirements of one of the following alternatives must be met:

(1) Time/temperature – an increased sewage biosolids temperature should be maintained for a prescribed period of time according to the Part 503 guidelines.

(2) Alkaline treatment – the pH of the biosolids is raised to greater than 12 for at least seventy-two hours.

(3) Prior testing for enteric viruses/viable helminth ova – if the biosolids is analyzed before the pathogen reduction process and found to have densities or enteric virus <1 plaque forming unit per 4 g TS and <1 viable helminth ova per 4 g TS, the biosolids is Class A with respect to enteric viruses and helminth ova until the next monitoring session.

(4) No prior testing for enteric viruses/viable helminth ova – if the biosolids is not analyzed before pathogen reduction processing for enteric viruses and viable helminth ova, the biosolids must meet the enteric virus and viable helminth ova levels listed below to be Class A at the time the biosolids is given away or sold as compost. *Note:* The density of viable helminth ova must be <1 per 4 g Ts. The density of enteric viruses must be <1 plaque forming unit (pfu)/4 g TS.

(5) The sewage biosolids must be treated by a Process to Reduce Further Pathogens (PRFP); that is, by composting. PFRP can be obtained by using either the in-vessel or the static aerated pile (ASP) composting methods. Using either of these methods, the temperature of the sewage biosolids must be maintained at 55°C or higher for three days (U.S. EPA 1993; Outwater, 1994).

For providing operational information on the ASP model composting facility used in this text, the *Salmonella* sp. density requirement and PFRP treatment process are used to meet the EPA's criteria for classifying biosolids as Class A.

VECTOR ATTRACTION REDUCTION REQUIREMENTS

The ASP model composting facility meets the vector attraction reduction requirements by utilizing the anaerobic digestion of biosolids alternative. (Other alternatives are available; they are listed in the Part 503 rule.)

Anaerobic digestion of sewage biosolids meets the vector attraction reduction requirements when the mass of volatile solids is reduced by 38% or more. Volatile solids reduction is measured between the raw biosolids before stabilization and the sewage biosolids ready for delivery to the composting site. An anaerobic digester that is well maintained and functioning as required will have little trouble meeting the requirements of this criterion (U.S. EPA 40 Part 503, 1993).

SAMPLING AND TESTING REQUIRMENTS (ASP MODEL)

The objective of sampling and testing biosolids-derived compost is to provide the necessary information to control unit process operation at the ASP model composting facility, which ensures compliance with regulatory standards to the fullest extent possible. Because the wastewater treatment and biosolids-derived composting process is a dynamic entity (changes with advances in technology), following a sound sampling and testing protocol is essential if the composting process is to be accomplished in line with the trend toward stricter regulatory requirements for processing compost for distribution. Moreover, a sound sampling and testing program can provide data that may assist future upgrading plans. From an economic point of view (operating and personnel costs), sampling and testing data may assist planners in determining the most cost-effective operating stratagem.

DATA RECORDING

Whenever sampling and testing is carried out, records must be kept (see Figure 11.1). From Figure 11.1 it can be seen that a series of entries are to be recorded related to the following:

(1) Sludge (biosolids type) % TS & VS
(2) Total sludge (biosolids) volume
(3) Bulking agent type % TS & VS
(4) Re-compost % TS & VS
(5) Compost mix % TS & VS
(6) Bulking agent to sludge (biosolids) ratio
(7) Other pertinent information including: daily monitoring requirements for
    - pile temp in °C
    - pile oxygen levels in %
    - blower operation cycles
    - blower pressures
    - trough numbers @ damper setting %
    - ambient temperature in °C
    - precipitation in inches
    - operator's initials

SAMPLING LOCATIONS

In previous sections, the sampling requirements related to biosolids processed at treatment plants and then loaded onto trucks for transport to the

PENINSULA COMPOSTING FACILITY
EXTENDED AERATED PILE DATA SHEET
HAMPTON ROADS SANITATION DISTRICT

[illegible]

PILE NUMBER ______________

INITIAL PILE CONSTRUCTION

SLUDGE TYPE	% TS	% VS	VOLUME YD3
A.			
B.			
C.			
TOTAL SLUDGE VOLUME			
BULKING AGENT TYPE			
A.			
B.			
TOTAL BULKING AGENT VOLUME			
RECOMPOST			
COMPOST MIX			
BULKING AGENT TO SLUDGE RATIO ____ : ____			
TOP BLANKET MATERIAL TYPE ______ HR IN ____			
TOE BLANKET MATERIAL TYPE ______ HR IN ____			
BLOWERS IN USE: TROUGHS IN USE:			
DATE PILE TAKEN DOWN			

TRANSPORTED TO	FINAL COMPOST	
____ DRYING ____ CURING ____ SCREENING ____ MIXING	% TS	% VS

PILE TEMPERATURE & O_2 SAMPLING LOCATIONS

DAILY MONITORING

DAY	PILE TEMP.-°C 1	2	3	4	5	6	PILE O_2 – % 2	4	6	BLOWER CYCLE ON/OFF	BLOWER PRESS (+)(−)	TROUGH NUMBER / DAMPER SETTING – %			AMB. TEMP. °C	PRECIP. INCHES	OPERATORS INITIALS
0																	
1																	
2																	
3																	
4																	
5																	
6																	
7																	
8																	
9																	
10																	
11																	
12																	
13																	
14																	
15																	
16																	
17																	
18																	
19																	
20																	
21																	

COMMENTS:

Figure 11.1 Daily operation record for HRSD composting facility.

TABLE 11.3. **Composting Facility Sample Locations.**

Sample	Location	Frequency	Type	Special Notes	Sampling Device
Mix (plant)	Drying building	Per load	Comp	1 Gallon	Quart jar
Mix (plant)	Drying building	Per load	Comp	1 Gallon	Quart jar
New WC	Drying building	Per load	Comp	1 Gallon	Quart jar
RC	Drying building	Per load	Comp	1 Gallon	Quart jar
Tear-down	Compost pad	21 days	Grab	1 Quart	Quart jar
Feed screen	Drying building	Each hour	Comp	1 Quart	Quart jar
Screen (fresh)	Storage pad	Once/monthly	Grab	Plastic bag	Plastic shovel
Screen (old)	Storage pad	Once/monthly	Grab	Plastic bag	Plastic shovel
Screen	Bagged compost	Once/monthly	Grab	40 lb bag	40 lb bag

Note: New WC = new woodchips. RC = recovered chips.
All routine samples are taken by metal shovel. All composite containers are plastic. All jars are made of glass. Screen compost is taken by HRSD technical services division for further analysis; they use a plastic bag and plastic shovel for sampling on metal analysis. On bagged compost, the whole 40-pound bag is taken.
Source: Sampling locations are listed for HRSD's Peninsula Compost Facility in Newport News, Virginia (Finley, 1996); used with permission.

composting facility were discussed. The main point to remember is that when the truck driver arrives at the composting facility, he/she must be able to inform the composting facility operator of the total solids percent concentration of the load being delivered. This information is vital for determining the correct ratio of bulking agent to biosolids to be used in the mixing process.

Along with the above-mentioned precomposting sampling requirements, other samples must be taken during the composting operation. To gain a better understanding of typical sample locations at a typical biosolids-derived composting facility, Table 11.3 is provided.

SAMPLE TYPES

In Table 11.3, one column for data entry is titled "TYPE." The "TYPE" column refers to grab or composite samples. Simply stated (with regard to composting), a grab sample is a portion of the biosolids-derived compost removed in a manner that enhances the probability that it is representative of the product at the instant it is taken (see Figure 11.2). Grab samples are required to establish the variability of the product with respect to time.

A composite sample is a mixture (combined in a single container) of equal-volume grab samples taken over a period of time, with the volume of the individual samples usually being proportional to the production flow of the product at the time the sample is taken (McGhee, 1991). Composite sampling can give a partial evaluation of the variability of compost composition with time.

To gain a better appreciation of how composite samples are taken in relation to the composting process, the following example is provided. *Note:* The following procedure is used by Hampton Roads Sanitation District for its Nutri-Green compost product produced at the Peninsula Composting Facility in Newport News, Virginia.

SAMPLE

Composite – Nutri-Green distribution pile compost.

EQUIPMENT

- quart jar with plastic top
- five-gallon plastic bucket
- plastic scoop
- preprinted laboratory label
- shovel

Figure 11.2 Compost facility operator taking a grab sample.

PROCEDURE

(1) Ensure all sampling equipment is clean, dry, and free from rust or dirt.
(2) Collect and composite ten consecutive representative grab samples of approximately equal size from the compost distribution pile as follows:
   - Randomly, select ten representative sampling locations at various heights and from all sides of the distribution pile.
   - Using the shovel, dig sample holes into the compost distribution pile at each sample location to depths alternating between 1 to 2 feet.
   - Collect one grab compost sample of approximately equal size from each of the ten sampling holes using the plastic scoop. Pour each grab sample into the clean plastic bucket.
   - Mix the bucket contents with the plastic scoop thoroughly after all ten samples have been collected.
(3) Fill the quart jar with the mixed contents of the sampling bucket. Discard the sample remaining in the bucket.
(4) Affix the preprinted laboratory label to the sample jar top, then date and sign the label.
(5) Refrigerate the sample between 1–4°C until the laboratory courier receives the sample.
(6) Rinse the five-gallon bucket and plastic scoop with potable water and shake dry.

SAMPLING PROCEDURES

After determining the location and type of compost sample to be taken, a composting facility sampling procedure must be developed. Each composting facility should develop sampling procedures tailored to its own particular operation. Sampling procedures should include detailed information on, for example, composite samples, composite factor, flow and flowrate, grab samples, labeling requirements, sample nomenclature, preparation for shipment to the testing laboratory, and compost sample procedures for *Salmonella* testing.

When information is being sought on sampling and testing procedures, the ultimate (definitive) reference is the current edition of *Standard Methods*. The sampling and testing procedures discussed in various sections of this text have been modified to fit a particular composting operation. However, some of the general material was obtained from the current edition of *Standard Methods,* and equipment operation procedures were obtained from the manufacturer's guidelines.

COMPOST TESTING

Along with various sampling requirements, certain tests are required to be performed at the composting facility (*Salmonella* is usually sampled and tested by qualified laboratory personnel), and the results are entered in the daily operating record (see Figure 11.1).

The following lists the various types of tests performed at typical composting facilities.

(1) Air velocity (air velocity meter) gaseous oxygen concentration – compost piles
(2) % moisture concentration (Denver Instruments, Ohaus balance, and centrifuge methods)
(3) % solids concentration (Denver Instruments, Ohaus balance, and centrifuge methods)
(4) pH
(5) Rainfall – rain gauge
(6) Temperature – ambient
(7) Temperature (temperature probes and meter)
(8) *Salmonella*

In the following sections, a detailed sample testing procedure including required testing equipment, is provided for each of the above tests.

AIR VELOCITY (AIR VELOCITY METER)

This test is conducted to ensure that the composting facility aeration system and associated equipment is operating as designed and within the parameters as set by the facility.

Equipment

The equipment used is a Weather Measure Model No. W241 Air Velocity Meter, air velocity probe.

Procedure

(1) Check the meter battery. *Note:* The batteries are charged satisfactorily when the meter indicator deflects to the "battery OK" range.
(2) Connect the probe to the air velocity meter.
(3) Take the meter and probe to the testing site.
(4) Expose the sensor from the shield of the probe a distance of one-half the diameter of the air discharge pipe to be tested. *Note:* Each etched line on the probe equals 1 inch.

(5) Insert the sensor into the air discharge pipe such that:
- The face of the shield of the probe is flush with the face of the air discharge pipe.
- The etched lines on the probe are facing into the air flow.
- The sensor is perpendicular to the air discharge pipe and air flow.

(6) Determine the proper air velocity scale. *Note:* Never peg the meter indicator at full scale to prevent damage to the meter. Withdraw the probe immediately from the air discharge pipe if the meter indicator pegs at full scale.
- Switch the meter function mode to the "6,000" scale.
- Proceed to Step 7 if the air velocity is above 1,000.
- Switch the meter function mode to the "1,250" scale if the air velocity is below 1,000 and proceed to Step 7.
- Switch the meter function mode to the "300" scale if the air velocity is below 300.

(7) Read the air velocity expressed as feet per minute (ft/min) using the selected air velocity scale when the meter indicator stops deflecting.

(8) Switch the meter function mode to the "off" position.

(9) Push the shield of the probe back into position fully covering the sensor.

(10) Disconnect the probe from the meter.

Note: When the air velocity test is conducted, ensure that the person conducting the test records the readings on the daily monitoring section of the data sheet. In this case, the daily reading would be recorded in the column labeled "BLOWER PRESS" (+) (−). If for some reason it becomes necessary to take and record several readings (e.g., for troubleshooting purposes), these readings should be recorded in the COMMENTS section of the data sheet.

GASEOUS OXYGEN CONCENTRATION—COMPOST PILES

Equipment

The equipment used is a Teledyne Model No. 320P Portable Oxygen meter, oxygen probe (see Figure 11.3).

Procedure

(1) Check the meter battery.
- Disconnect the battery charger.
- Switch the meter function mode to the "battery test" position.

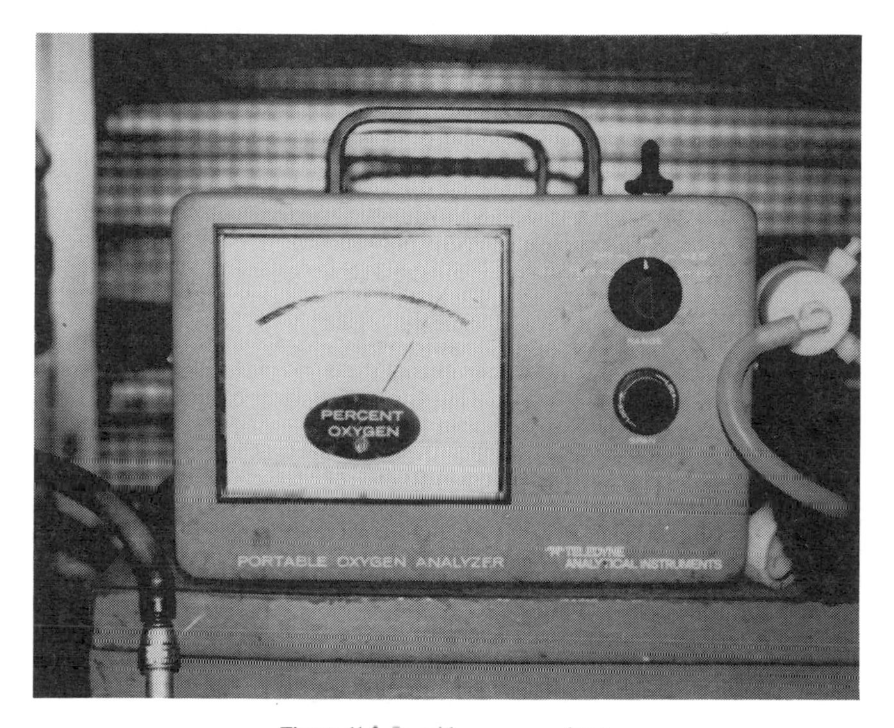

Figure 11.3 Portable oxygen analyzer.

- Read the battery charge using the battery scale when the meter indicator stops deflecting. *Note:* The batteries are charged satisfactorily when the meter indicator deflects to the "battery limits" range.

(2) Check the oxygen meter water trap (see Figures 11.4 and 11.5).
- Determine the color of the silicagel crystals in the water trap. *Note:* The silicagel crystals must be blue for the water trap to function properly. (1) Proceed to Step 3 if the silicagel crystals are blue or (2) proceed to the following if the silicagel crystals are pink.
- Remove the rubber stopper from the water trap.
- Empty out the pink silicagel crystals and remove the wool filter.
- Place a new wool filter in the bottom of the water trap and fill the water trap with new blue silicagel crystals.
- Reinsert the rubber stopper in the water trap.

(3) Clean any clogged air holes on the oxygen probe.

(4) Remove the fuel cell saver cap from the meter and replace with the flowthrough adapter.

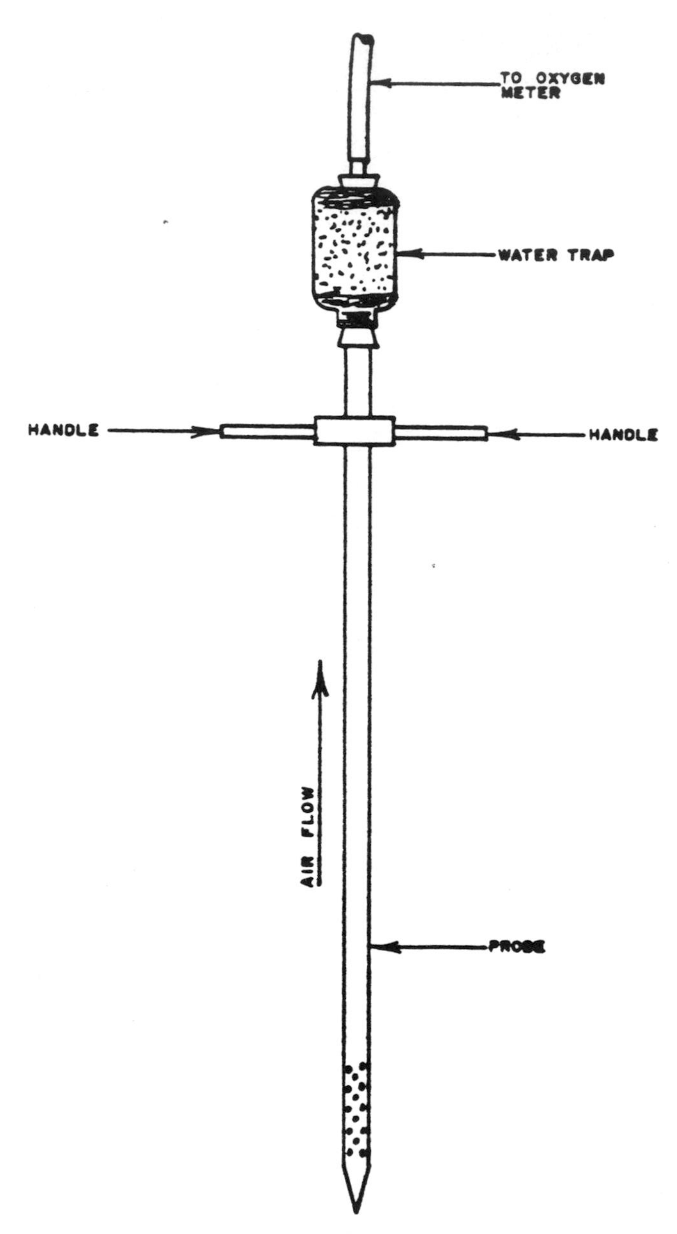

Figure 11.4 Oxygen probe used by HRSD composting facility when testing for oxygen level; used with permission.

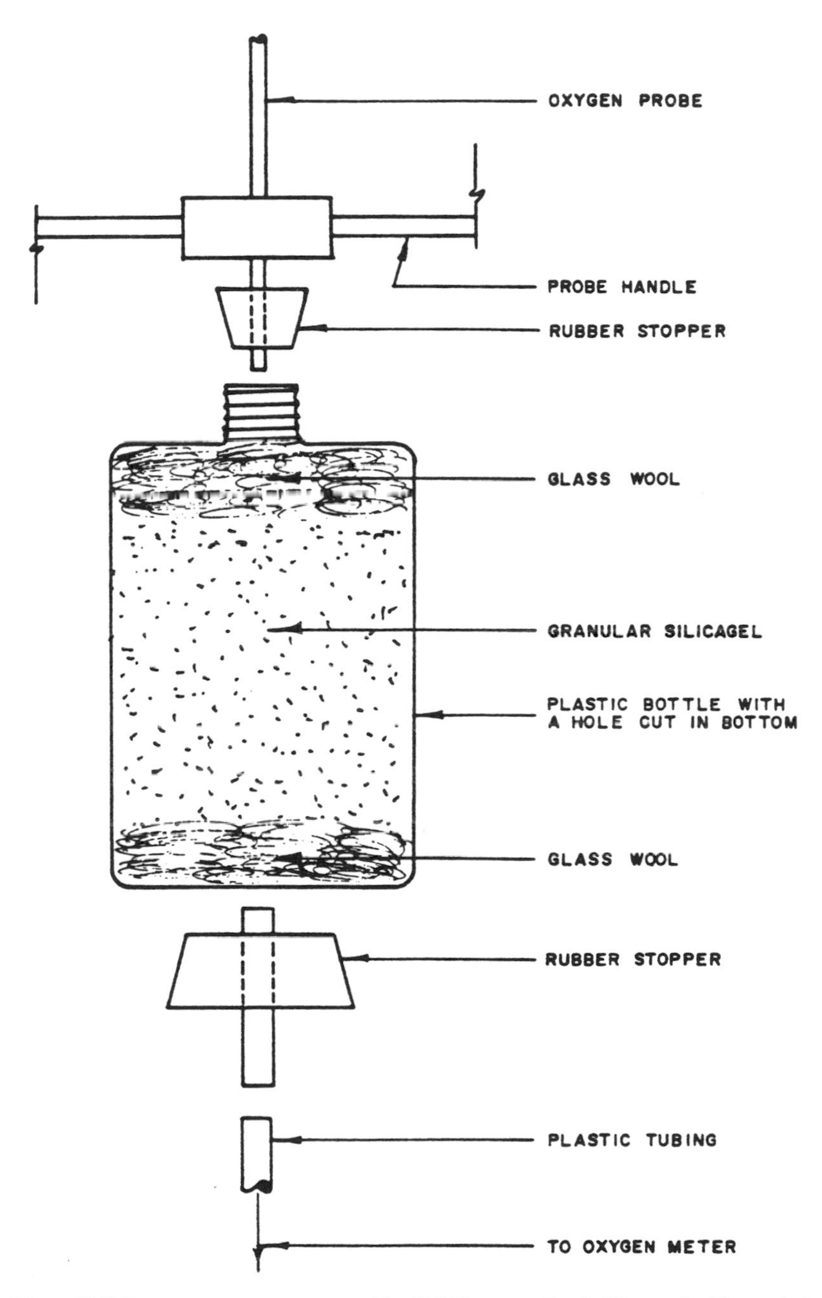

Figure 11.5 Oxygen meter water trap used by HRSD composting facility; used with permission.

(5) Connect the plastic tubing of the probe to the meter.
(6) Calibrate the oxygen meter in accordance with the manufacturer's directions.
(7) Take the meter and probe to the testing site in the field.
(8) Insert the probe into the compost pile to the required depth.
(9) Determine the proper gaseous oxygen concentration range.
- Determine if the gaseous oxygen concentration is above 10%. (1) Switch the meter function mode to the "high" range. (2) Proceed to Step 10 if the gaseous oxygen concentration is above 10%, or (3) proceed to the following if the gaseous oxygen concentration is below 10%.
- Determine if the gaseous oxygen concentration is between 5 to 10%. (1) Switch the meter function mode to the "medium" scale. (2) Proceed to Step 10 if the gaseous oxygen concentration is between 5 to 10%, or (3) switch the meter function mode to the "low" scale if the gaseous oxygen concentration is below 5%.

(10) Read the gaseous oxygen concentration expressed as percent using the selected oxygen concentration scale when the meter indicator stops deflecting. Record readings on the daily monitoring section: PILE OXYGEN %.
(11) Switch the meter function mode to the "off" position.
(12) Withdraw the probe from the compost pile.
(13) Repeat Step 8 through 12 for additional testing sites as appropriate.
(14) Disconnect the plastic tubing of the probe from the meter.
(15) Remove the flowthrough adaptor from the meter and replace with the fuel cell saver cap.

PERCENT MOISTURE AND SOLIDS CONCENTRATION

For these two tests, follow the procedures listed in Chapter 2: Denver Instruments moisture analyzer method, laboratory centrifuge method, and/or Ohaus balance method. Ensure that all findings are properly recorded.

pH TEST

For the compost pH test, several testing devices are available. Again, the main concern is that findings are recorded in the appropriate record or log sheets.

RAINFALL—RAIN GAUGE

It is important to measure any precipitation that falls on the composting facility. Precipitation can be measured quite easily by using a standard rain gauge. Again, all readings must be recorded in the appropriate records and logs.

TEMPERATURE—AMBIENT

The procedure for measuring compost site ambient temperature is as follows.

Equipment

The equipment used is a mercury-filled thermometer graduated in degrees Centigrade.

Procedure

(1) Position the thermometer in the field such that:
- It is not located in direct sunlight.
- It is not located in abnormal wind currents.

(2) Wait a few moments until the mercury column stops deflecting at a fast rate.

(3) Read the temperature expressed as degrees Centigrade using the top of the mercury column as the reference point. Ensure that readings are recorded on the appropriate record or log sheet.

TEMPERATURE—COMPOST PILES (TEMPERATURE PROBE METHOD)

Equipment

The equipment used is an Atkins Model No. 44011-C Portable Temperature Meter, temperature probe.

Procedure

(1) Check the meter battery charge.
- Disconnect the battery charger.
- Press the meter operating button.

- Read the battery charge using the meter display. *Note:* The batteries are charged satisfactorily when the meter display does not indicate "low battery."
- Proceed to Step 2 if the meter display does not indicate "low battery."
- Recharge the batteries by connecting the meter to the battery charger if the meter display indicates "low battery." *Note:* Do not recharge the batteries continuously for longer than 24 hours. A 12- to 16-hour battery recharging should last for several months.

(2) Connect the cable of the probe to the temperature meter.
(3) Take the meter and probe to the testing site in the field.
(4) Insert the probe into compost pile to the required depth.
(5) Turn on the meter by pressing the meter operating button.
(6) Read the temperature expressed as degrees Centigrade on the meter display when the digital readout stops changing.
(7) Turn off the meter by releasing the meter operating button.
(8) Withdraw the probe from the compost pile.
(9) Repeat Steps 4 through 7 for additional testing sites as appropriate.
(10) Disconnect the probe from the meter.
(11) Wipe down the probe and tip.

SALMONELLA

As stated earlier, EPA's Part 503 rule requires that wastewater biosolids meet certain criteria in order to be classified as "Exceptional Quality." One of the tests that can be conducted on the compost product is the *Salmonella* test. If the compost shows no problem with *Salmonella* contamination and the heavy metal content is within prescribed limits (see Table 11.2), the compost product (which meets the Process for Further Reduction of Pathogens [PFRP] requirements because of the composting process) is considered safe for sale to the public for use in land appliction. A procedure used in the ASP model composting facility for *Salmonella* sampling is provided below.

Equipment

The equipment used is a sterile container (glass quart, or whirl-pak 27-ounce bag), shovel.

Procedure

(1) Designate one day each month for collecting the samples. *Note:* For quality control/quality analysis (QA/QC) purposes, the treatment plant lab should be responsible for taking the samples and designating the day for collecting them.
(2) Randomly select a sample location from the cured screened compost on site. If no cured screened compost is available, select a sample location with cured unscreened compost.
(3) Clean shovel with potable water.
(4) Dig approximately 18 inches into the center of the selected sample location with the clean shovel.
(5) Fill a sterile sample container with compost from this interior location of the pile.
(6) Randomly select a second sample location as in Step 2.
(7) Repeat Steps 3, 4, and 5.
(8) Immediately take both samples to the testing laboratory for analysis.
(9) Ensure laboratory results are recorded in the COMMENTS sections of the compost facility daily operating log.

SAMPLE LABELING

Compost operators who take samples for analysis waste a lot of time and money if they do not properly label the samples. In order for proper analysis to be performed and entries recorded, each compost sample must be correctly labeled. All samples must be identified with routine nomenclature and sample data. If preprinted labels are not available, a handwritten label must include the following information: site name, sample name, sample code (if used), sample data, and operator's signature. *Note:* When samples are to be shipped to the treatment plant laboratory for analysis, not only must they be clearly labeled, but the labels must also be securely attached to the sample container.

CHAPTER 12

Odor Control

> When a backyard cookout is canceled because of a local malodor, or when the homeowner feels he must close his windows and install air purifiers, or when he operates his air-conditioning system when outside temperatures do not require air cooling, these behaviors may be translated into dollar costs. In fact, the courts often recognize such actions as evidence that odorous emissions are damaging and that compensation should be made by the offender. (Cheremisinoff & Young, 1981)

INTRODUCTION

IN the preceding statement, Cheremisinoff and Young point out (with regard to the interface between the public and industry) a well-known fact: When impairment of use and enjoyment of property results from the operation of an industrial process, such as composting, serious complaints from political leaders, community administrators, and the public are almost guaranteed to occur.

Impairment of use and enjoyment of property is only one of the effects associated with the industrial production of nuisance conditions, such as malodors. As an example, consider the effects of malodors with respect to dollar costs: (1) When malodors pervade a community or industrial setting, existing property may decrease in value. (2) A dollar cost may be involved with the loss of use of surrounding open land; that is, some areas surrounding industrial complexes are designated for recreational use, but how many people are going to seek recreation in an area where the odor is offensive? (3) The direct personal effect of offensive odor production is another factor that must be considered. For example, it is not unusual for people to complain that malodors cause them to experience a loss of sleep, loss of appetite, and nausea. When the public begins to complain, and when their complaints turn into law suits, obviously, the industrial com-

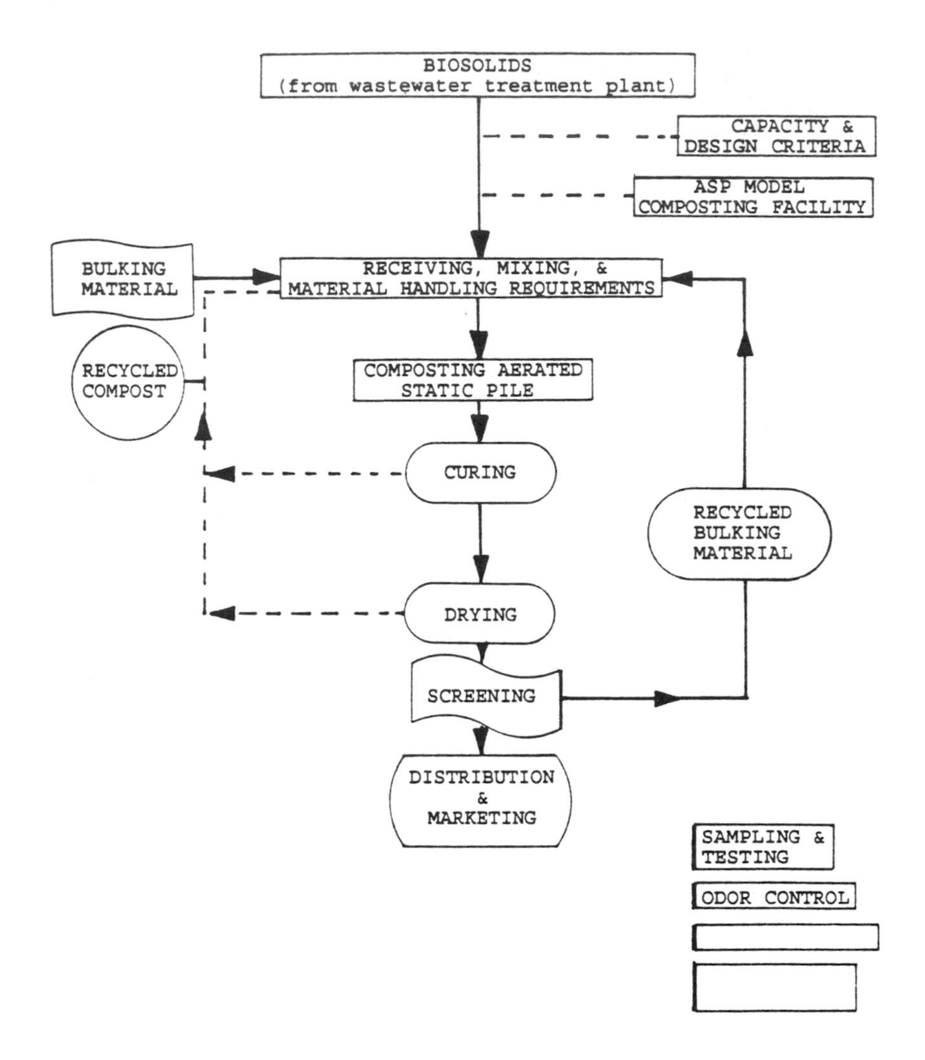

plex responsible for the production of the offensive odors can be put in a serious financial bind.

How do industrial managers prevent such serious repercussions? Finstein et al. (1986) may have the answer: When a specific, objective criterion of process performance is instituted, unnecessary expense and widespread problems, particularly related to the generation of noxious odors, can be avoided. This chapter presents a specific, objective criterion of process performance that can prevent unnecessary expense and widespread problems with odor generation from biosolids-derived composting processes. *Note:* Many of the concepts presented in the following section are the result of years of research on odor control at wastewater treatment plant, pumping stations, and a composting facility. Most of these studies were conducted by HRSD project engineer, Brian McNamara (1996), with critical analysis (fine-tuning) supplied by G. David Waltrip (1996), director of treatment, HRSD.

COMPOSTING AND ODOR CONTROL

Since the major operational problem associated with composting is reportedly the production of odors, composting-process odor control must receive serious consideration (McGhee, 1991). Planners, designers, decision-makers and engineers responsible for planning, designing, engineering, and funding composting facilities generally take into account the need to "site" the proposed composting facility in a location that is suitable for its intended operation.

Suitability of location is important for several reasons. For example, when a compost site is in the planning stages, consideration must be given to accessibility: Can the truckloads of biosolids cake be delivered economically to the proposed site? Are the roads suitable for small private vehicles as well as large Ram-E-Jec-type trucks and trailers? Does the proposed site afford enough room for the entire composting operation? Can bulking agents be easily obtained, transported, and stored on site? Is there enough surrounding land to form an adequate buffer zone around the site?

As stated, when potential odor problems become a planning, designing, and engineering concern (as they should be in the biosolids composting process), conventional practice calls for inclusion of an adequate buffer zone. What exactly is an "adequate" buffer zone?

In years past, when cities were smaller, finding a composting site with a large enough natural buffer zone was not difficult. Thus, it was not unusual to build a composting facility in an area where few people lived and where hundreds of undeveloped acres of forest and wide open spaces were the norm. With the passage of time, however, the population and suburbia grew in all directions. What might have been an isolated composting facility with an extensive buffer zone twenty years ago today is surrounded on all sides by urban sprawl.

The point is that in the past when composting facilities were built, in most cases, little thought was given to future growth near or around them. Thus, the composting facility had little to worry about in terms of odor complaints from the public. Today, this is no longer the case.

In addressing odor control problems with biosolids-derived composting facilities, two main scenarios are addressed in this text: new construction and established facilities.

NEW CONSTRUCTION

In new construction, the planners, designers, and engineers have the luxury of basing their plans, designs, and processes on data (lessons learned) derived from other composting operations. This is a huge advantage in that for the past twenty-five years several biosolids-derived composting operations have been in operation throughout the United States. Data obtained from these operations can be used to select the proper process, design the proper facility, and properly train managers in site operations—all of which may result in an operation free of malodors (Corbitt, 1990).

However, new construction does face a major hurdle not normally experienced by older operations: siting. As pointed out earlier, the potential problems of finding a suitable site with an accompanying buffer zone are very real and troublesome. The never-ending encroachment of population and industry into what originally was "virgin" land space is a reality. The Not In My Backyard (NIMBY) syndrome is real. This is especially the case when the public has the perception that the new site will not be free of malodors. If a community perceives the siting of a compost facility in its "backyard" as a potential nuisance that will pollute the environment, decrease land values, and affect the quality of life, then the planners, designers, and engineers may be up against overwhelming opposition (Outwater, 1994). Even when planners, designers, and engineers openly and honestly present their plans, designs, and processes to the citizens, the struggle to get them to buy in on having a composting facility as a neighbor is not to be underestimated.

ESTABLISHED FACILITIES

According to an old saying, odors are not a problem until the neighbors complain. Long-established composting facilities may have enjoyed, at earlier times, an absence of neighbors and thus a lack of public awareness of their facility and the odors generated. In most cases, this is no longer the case, however. As stated earlier, massive expansion of urban development has encroached upon what used to be remote areas around most composting facilities. It is probably safe to say that if the local comunity has expanded into and become a close neighbor of an existing compost facility, and if the composting process is not carefully managed (controlled), odors can be a definite problem (Tchobanoglous et al., 1993); that is, the neighbors will complain.

THE NATURE OF ODORS

Attempting to get people to agree on the desirability or undesirability of an odor is difficult. Perception of odor and how well it is received by people has lot to do with association of the odors with their sources (Vesilind, 1980). As a case in point, consider the following example.

When a person, for the first time—not knowing the source—detects the heavy, earthy smell of biosolids-derived compost, he or she may like it or dislike it. The point is that each person perceives odor differently. Those who are not offended by the odor of composting biosolids may not give it a second thought. However, a few days or weeks later, if this same person is driving to work with a neighbor who is familiar with the source of that heavy, earthy compost odor and passes this information on to the unknowing person, his or her perception is likely to change. This phenomenon should not come as a surprise, considering that most individuals do not associate flushing their toilets with wastewater treatment and its ancillary processes (composting). However, when finding out, like the unknowing person in the car, people may perceive the heavy, earthy biosolids-derived compost odor in a very different way.

Along with knowledge of its source, odor has another characteristic that may impact people: the individual's sensitivity. As far as sensitivity is concerned, the key point to remember is that odor is subjective; that is, what is offensive to one person may not be to another (Outwater, 1994).

When attempting to address the factors that characterize an odor, it is wise to refer to the four independent factors described by Metcalf and Eddy (1991): "intensity, character, hedonics, and detectability" (p. 58).

Intensity refers to perceived strength of the odor as measured by an olfactometer. The character of an odor refers to any mental associations made by the subject sensing the odor. Hedonics refers to the relative degree of pleasantness or unpleasantness perceived by the subject, which is usually determined "by using a scale estimating the magnitudes of the aesthetic qualities found in odors" (Lue-Hing et al., 1992, p. 197). Finally, odor detectability refers to the number of dilutions that are needed to reduce an odor to its minimum detection point.

MEASURING ODORS

In measuring odors, the panel or dilution methods are usually used. The dilution method is used in the water treatment process to detect odors in water and, therefore, will not be discussed here. The panel method involves using ten or more people who make a judgment about the odor. Their individual judgments are subsequently recorded and analyzed. According to Vesilind (1980), the results of the panel method can be used to determine an "average opinion of the strength and nuisance value" of certain odors (p. 42).

When the panel method is used to measure odor, the parameter normally used for detecting odor is expressed as the number of effective dilutions-50 (ED-50). ED-50 is the number of fresh air dilutions required to reduce the odor level of a sample so that only 50% of the panel can smell it. Odor standards are based on odor control parameters such as ED-50, ED-10, ED-5, and others.

Note: In setting the odor control parameter for a new or an existing facility, a setpoint of ED-50 is not practical, since the odor level would be reduced (diluted) for only 50% of those who live near or come in close proximity to the compost facility. For this reason, ED-5 was created, whereby only 5% of the exposed subjects would perceive the composting odors (Wilber & Murray, 1990).

MALODOROUS COMPOUNDS IN BIOSOLIDS-DERIVED COMPOST

To characterize composting as a smelly process is correct. Whether or not this smelly process is offensive to the subject is another issue, depending, almost entirely, on individual sensitivity.

It is interesting to note that the ingredients important to the composting process all smell. These smelly but important ingredients include: amines, aromatics, terpenes, organic and inorganic sulfur, and fatty acids.

Generally associated with fats and oils-based industrial operations, amines are more commonly known for their distinct fishy odor. In composting, amines are a by-product of microbial decomposition and generally form during anaerobic fermentation. Aromatics are usually volatilized during aeration. When woodchips are used as the bulking agent in the biosolids mix, aromatics are produced during the aerobic composting as the lignins (in woodchips) break down. Likewise, terpenes (which are products of wood) are also present in compost piles that use woodchips as the bulking agent.

Probably, most wastewater specialists have been exposed to hydrogen sulfide, an inorganic sulfur compound, and its characteristic rotten egg odor. Under normal circumstances, when biosolids are received at the composting site, any hydrogen sulfide emissions are quickly reduced when the biosolids and bulking agent are mixed and formed into aerobic piles. However, there can be a problem with hydrogen sulfide emissions, if the mix is incorrect or if the biosolids is too wet. When the biosolids is wet, it tends to form into clumps, which can become anaerobic and will subsequently form and release hydrogen sulfide.

Whether described as "stinking like a skunk" or smelling like "decayed cabbage," organic sulfurs are generally present in all biosolids-derived composting piles. Of the various organic sulfur compounds found in compost piles, probably the best known is the methyl mercaptans (smells like decayed cabbage).

Fatty acids are generally produced under anaerobic conditions and do not add to odor generation problems unless the pile is allowed to go anaerobic.

In a recent study conducted by Muirhead et al. (1993a) at a wastewater treatment plant that included a composting facility, odor sources and odorous compounds from various treatment processes were differentiated as show in in Figure 12.1.

At a biosolids composting facility, any sensible odor-control management plan takes into account all the areas and components of the composting process that might cause odors. Most odor problems are generated in the composting and curing process air systems. However, at enclosed composting operations, odors from ancillary processes within the enclosure must also be taken into account. Enclosed systems must have a way to control or scrub air flow within the structure prior to its release to the outside environment.

In addition, usually, not all process areas at a composting facility are enclosed. These open areas (e.g., biosolids handling and mixing areas) can also cause odor control problems.

Building-ventilation air dispersion for enclosed facilities, biosolids handling and mixing areas, and composting and curing process air odor control measures will be addressed in the following sections.

BUILDING-VENTILATION AIR DISPERSION

Normally, fugitive odor emissions from the compost and curing piles, from cured compost screening, and from screened compost storage piles

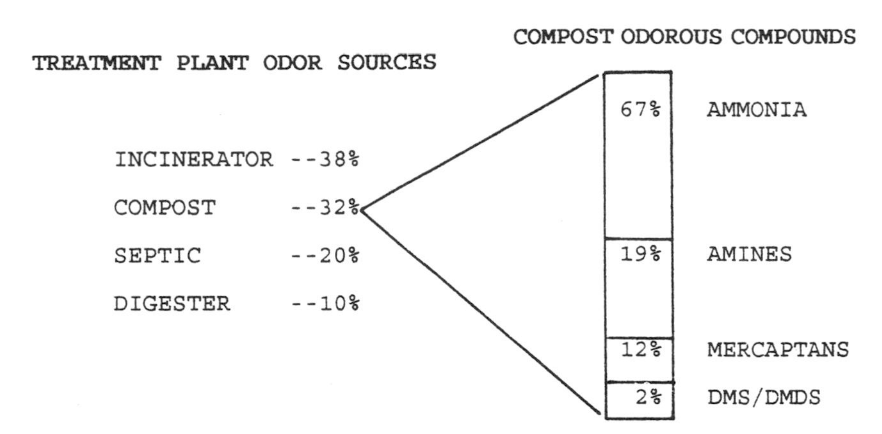

Figure 12.1 Treatment plant odor sources and odorous compounds from composting facility. Source: Adapted from Muirhead et al. (1993a), p. 67.

are relatively low. However, these areas should be enclosed for odor containment and controlled discharge. To gain a better perspective on odor-control design considerations for various composting processes, the odor-control and dispersion processes used by the ASP model composting facility and others will be discussed.

Odor-control and dispersion modeling analyses prepared for the ASP model composting facility pointed to the need for the buildings to be continuously ventilated at up to twelve air changes per hour (ach). Twelve (ach) will allow for the safe operation of internal combustion engine-driven equipment within the enclosed areas, while also preventing accumulation of heat and odors. The odor-dispersion modeling part of the analyses concluded that building ventilation air could be discharged directly to the atmosphere through high-velocity roof-mounted fans without significant effects on offsite odors.

As part of the odor-control renovation project study conducted for the ASP model facility, it was recommended that the compost mixing operations, compost and curing piles, recycled woodchip storage and woodchip screening operations be enclosed in one building, with finished compost storage piles enclosed in a separate building.

Management followed the recommendation, implementing it by building (a) compost process building, 500′ × 300′ × 25′ clear height; and (b) compost storage building, 270′ × 190′ × 25′ clear height. At a ventilation rate of twelve air changes per hour (ach), the compost process building ventilation rate is 760,000 cfm, whereas the compost storage building ventilation rate is 230,000 cfm. Separate ventilation systems were used for each building. The systems each use two high-velocity cooling tower fans mounted on a raised platform with a discharge shroud extending slightly above roof level. If better dispersion is required, the building air can be discharged through taller stacks (HRSD/Black & Veatch, 1993).

ODOR MANAGEMENT: BIOSOLIDS AND MIXING AREA

When considering improvements in odor control for composting facilities, dewatered biosolids receiving and handling operations areas should not be overlooked. If mechanical mixing is used, it is prudent to ventilate the biosolids receiving bins, conveyors and mixers directly and convey the ventilation air to the process air scrubbers for treatment. For the ASP model, the volume of air that is to be treated is a little less than 2,000 cfm.

In addition, when front-end loader mixing is used, it is necessary to enclose the mixing area separately and treat the ventilation air. For the ASP model, where front-end loaders are used in the mixing process, the volume of air required is approximately 130,000 cfm. Attempting to treat 130,000 cfm of air in a wet scrubber system is cost prohibitive. A more cost-effective way to reduce odor in the mixing area ventilation air is to pass it through a biofilter.

Biofilters are different from compost scrubbers in that the adsorption media used for the biofilter bed is a layer of well-ventilated and biologically cultivated soil or deodorized finished compost. There are drawbacks involved with using biofilters. First, since biofilters usually require large land requirements, their use is generally limited to small plants (Lue-Hing et al., 1993). Second, it is difficult to recapture the treated air for dispersion through a discharge stack (HRSD/Black & Veatch, 1993). To determine whether untreated air for discharge through a dispersion stack or biofilter-treated air discharged at ground level is more appropriate for a particular site, a pilot study and associated testing should be conducted.

For purposes of illustration, if the ASP model was renovated to include mechanical mixing of biosolids and woodchips, 2,000 cfm of ventilation air from the biosolids receiving bins, covered conveyors and mixers is to be conveyed to the process air scrubbers where it is combined with the 35,000 cfm of process air for treatment, This renovation will require that the capacity of the process air scrubbers system be increased to 37,000 cfm.

COMPOSTING PROCESS AIR ODOR MANAGEMENT

In attempting to minimize production of odors in the biosolids-derived composting process, both proper process design and operation are critical. Process equipment that satisfies aeration requirements, proper temperature control, and mixing must receive special attention. However, having properly designed equipment on site is not the complete answer to solving potential odor problems. In addition to properly designed composting equipment and monitoring devices, it is necessary to ensure that standard operating procedures are established and that operators are trained on and required to follow these procedures.

Even with a properly designed composting process, with properly installed associated monitoring equipment, and with composting operations personnel who are competent at their jobs, the biosolids-derived composting process will produce odors. Remember, odor is not a problem until the neighbors complain. However, the neighbors will complain if odors from composting operations are not managed, contained, controlled, and minimized.

The solution to the odor control problem presented in this text is modeled after conditions that could be found at the ASP model site, discounting the odor-control renovations mentioned earlier. That is, the ASP model composting facility, without renovation, is an open operation. With the exception of the shed-type structure, the process is not enclosed. In composting, whether the process is enclosed or open, odor "events" are inevitable.

ODOR CONTROL

For illustrative purposes, assume that neighbors of the ASP model composting facility have consistently complained about the odors emanating from the facility. The facility manager has decided to take remedial action and has hired an engineering consulting firm to determine the best method for solving the problem.

The engineering consulting firm recommends that the most practical solution is to totally enclose the composting operation, noting that with the composting process totally enclosed, odors will be contained. The problem then becomes: What is to be done with the odors that are contained? According to Tchobanoglous et al. (1993), when a facility is enclosed, odor-control equipment "such as packed towers, spray towers, activated carbon contractors, biological filters, and compost filters have been used for odor management" (p. 316).

For odor management at the ASP model composting facility, the engineering consulting firm recommended that some type of wet scrubbing system be used after the process is enclosed. To decide what type of system would be best suited for the facility, the engineering consultant reviewed data from an odor-control study conducted by Hampton Roads Sanitation District (HRSD) for its composting facility.

The original 1990 report recommended two stages of wet scrubbing for process air from both the composting and the curing piles. The recommendation required aeration blowers to draw air into the piles so that it could be captured for scrubbing. An additional scrubbing tower was planned as a standby unit to ensure that two scrubbers would always be available for continuous operation. It was also suggested that the third scrubber be placed in operation as a third-stage unit to provide additional reduction if necessary.

The engineering consultant noted that HRSD had indicated that experience at other biosolids composting facilities since 1990 had shown that three stages of scrubbing are needed for effective odor removal. In addition, the consultant determined that moisture removal from the hot, saturated process air is required ahead of the chemical scrubbers to prevent heat build-up and dilution of chemical solutions used in the scrubbers.

The next step the engineering consultant took was to make a comparison between odor-control systems at various locations. System types included the mist and the packed tower scrubbers. Table 12.1 shows the data obtained from this comparison.

ODOR-CONTROL SCRUBBER ALTERNATIVES

As a result of preliminary investigation (as related earlier) for odor control in the ASP model composting facility and of the odor-control design steps recommended by Metcalf and Eddy (1991) (see Table 12.2), the design engineers recommended installation of a scrubber system. Two types of wet scrubbing systems, a three stage mist system and a three-stage packed tower system, were evaluated for treating process air for odor removal.

In the following sections, these two systems are described and compared on the basis of performance, cost, reliability, operation and maintenance. Moreover, preliminary system designs are also presented, based on the manufacturer's recommendations with minor modifications to make the systems more comparable.

Mist Scrubber System

A mist scrubber is a gas-liquid contacting system that consists of a contact chamber and atomizing nozzles for injecting chemical solutions. Mist scrubbers produce a dense fog of fine droplets (5–20 microns in diameter) to maximize the liquid surface area for contact with the airstream being treated. Air moves through the contact chamber co-current with the chemical feed solution. Compressed air is required to operate the atomizing nozzles. Water for makeup of chemical feed solutions must be filtered and softened to prevent scaling and clogging of the nozzles. The chemical solution is not normally recirculated, and any residual solution is discharged after a single pass through the contact chamber. The contact chamber drain is continuously monitored with a pH sensor, and the rate of chemical injection is automatically adjusted to maintain optimum pH for the treatment objective.

Manufacturers of mist systems with experience in municipal biosolids composting include Quad Technologies Inc. and Calvert Environmental Equipment Company. Information from Quad was used for this comparison (evaluation) because of their involvement in the development of a three-stage scrubbing process specifically designed to optimize odor removal from biosolids composting process air.

Mist Scrubber Installations

Extensive research was conducted at the Washington Suburban Sanitary

TABLE 12.1. Odor-Control System Comparison.

System Type	Installations	Performance	Proposed System Description
Mist scrubber	Montgomery County, MD 40 dtpd aerated static pile Odor-Control System: 1—H_2SO_4 2—H_2SO_4, NaOCl, Tide, pH 8.0 3—NaOH, peroxide 500,000 dilution, 110 ft stack	NH4 99% in—90 ppm DMDS 90% removal 90% of time Odor 80% inlet—500 ED-50, outlet—90 ED-50	3-stage, 30,000 cfm 4 fiberglass chambers—12′ dia × 48′ tall Reaction time—7 seconds 1—H_2SO_4 2&3—H_2SO_4, NaCOl, Tide, pH 6.5 4—NaOH, H_2O_2, pH 8 100,000 cfm dilution fan 100′ stack, 3′ dia. Water softener Air comprerssor Redundant nozzles, feed pumps & fans Four 5,000 gal stor. tanks w/containment
	Lancaster, PA 32 dtpd in-vessel Odor-Control System: Quad, 3-stage, 15,000 fcm 1—HCl, pH 7, 450 gpm water 2—HCl, NaOCl, Tide, pH 6.5 2 outlet—NaOH, peroxide 40,000 cfm dilution, 42′ stack	Generally good. ''Burnt chemical'' odor w/pH 4.3 in 2nd stage. Odor peaks not treated at vessel startup.	

TABLE 12.1. (continued).

System Type	Installations	Performance	Proposed System Description
Packed tower	Cape May County, NJ 20 dtpd in-vessel Odor-Control System: 3-stage, 20,000 cfm 1—H_2SO_4, ph 2.5–3.0, 700 gpm effluent, 350 recirc. 2—H_2SO_4, NaOCl, Tide, pH 6.5 3—NaOH, peroxide, pH 8.5, no dilution, 100′ stack	Informal sniff tests show good odor control. Plant effluent cools gas from 130 to 80° F. Cl_2 monitor controls outlet to 5 ppm.	4-stage, 30,000 cfm 4 fiberglass chambers—10′dia × 30′ tall, 14–16 packing depth Reaction time—7 seconds 1—Cooling tower with H_2SO_4 2—H_2SO_4 3 & 4—NaOH, NaOCl, pH 10 100,000 cfm dilution fan, 100 ft stack, 3′ diameter. Redundant recirc. pumps, chemical feed pumps, and fans. Four 5,000 gal. storage tanks w/containment to permit WSSC process
	Schenectady, NY in-vessel Odor-Control System: 2-stage, 46,000 cfm 2 parallel systems Pre 1—Quench tower, pH 1—H_2SO_4, pH 2–2.5, horiz. cross-flow scrubbers 2—H_2SO_4, NaOCl, pH 6.5–6.9, vertical towers, 10′ media No dilution, 2-50′ stacks 7.8 ac/hr building ventilation	NH4 >99.9%, inlet—300 ppm Amines >99.9%, inlet—20 ppm *DMDS 99%, inlet—15 ppm Odor >96%, inlet—558 ED-50, outlet—21 ED-50 Wood ash used pile odor	

*DMDS or dimethyl disulfide was identified by the Washington Suburban Sanitary Commission (WSSC) as the primary odorant in air from the Montgomery County Regional Composting Facility.

Source: Adaption from pilot study conducted by Hampton Roads Sanitation District and Black & Veatch (1993); used with permission.

TABLE 12.2. Steps in Designing a Wet Scrubber System.

1. Determine the characteristics and volumes of gas to be treated.
2. Define the exhaust requirements for the treated gas.
3. Select a scrubbing liquid based on the chemical nature and concentration of the odorous compounds to be removed.
4. Conduct pilot tests to determine design criteria and performance.

Source: Adapted from Metcalf & Eddy (1991).

Commission (WSSC) Site 2 composting facility in Montgomery County, Maryland, on a three-stage mist scrubber system. Several articles have been published pointing to the success of this odor-control systems (Goldstein, 1993). More specifically, the research effort at the WSSC site led to the development and refinement of an odor-control system that is quite effective.

Mist Scrubber Operation (WSSC Model)

The first-stage scrubber removes ammonia and other organic compounds by using an unscented commercial detergent (surfactant–surface active agent) containing a solution of sulfuric acid. *Note:* Surfactants help increase the removal of organic compounds by solubilizing organic compounds at the air-water interface by reducing surface tension (WEF & ASCE, 1995).

A series of fine and coarse spray nozzles, along with atomizing nozzles, are used in the first-stage scrubber to help cool the process air, which aids in coalescing the fine droplets of ammonia and organic compound-laden chemical solution in the airstream. Hot process air coming from the composting piles contains a high level of moisture, which significantly increases the liquid solution captured in the first-stage scrubber sump. In order to remove ammonia, organic compounds and excess heat, a large portion of the sump liquor is blown down to the drain level. From drains the sump solution is pumped along with fresh makeup liquid to the secondary coarse and fine nozzles.

In the second-stage scrubber, sodium hypochlorite and sulfuric acid are added, and the pH is maintained at approximately 6.5 to oxidize DMDS (dimethyl disulfide) and other reduced sulfur compounds. Finally, in the third-stage scrubber, hydrogen peroxide and sodium hydroxide are used at a pH of 8.0 for additional oxidation of organic compounds and to remove excess chlorine carryover from the second-stage scrubber.

Mist Scrubber Performance (WSSC Model)

Extensive documentation has been compiled by WSSC on the performance of the three-stage mist scrubber system at its Site 2 composting facility. After pilot scale testing of various scrubber types and system configurations, the original mist scrubber system was installed in 1985.

In this system, sulfuric acid was injected into the riser duct ahead of the first-stage contact chamber to reduce ammonia concentration in the process gases. To remove hydrogen sulfide and other oxidizable compounds, sodium hypochlorite was added to the first contact tower. Then, in the second tower, sulfuric acid was added to remove additional ammonia. A 500,000 cfm fan was used to dilute treated process air with fresh air in a 110-foot tall discharge stack. Even though the system provided significant odor removal, operators were unable to maintain adequate performance on a consistent basis.

After additional testing, the original mist scrubber system was modified to improve performance. Initial problems with inadequate first-stage ammonia removal and acid carryover from the first-stage riser duct were resolved by adding secondary coarse and fine spray nozzles, a pH control system, recirculation pump, and a sump. These changes converted the acid pretreatment step into a separate first-stage scrubber. Using gas chromatograph/mass spectrometry (GC/MS) analysis and odor panel testing of process gases and scrubber exhaust, DMDS (dimethyl-disulfide) and a class of organic terpenes were identified as the principal sources of odor remaining in the treated process gases. Subsequent laboratory testing confirmed that sodium hypochlorite was the most effective oxidant for DMDS removal; it was also found to be most effective at high concentrations and a pH of 6.5.

Full-scale testing conducted later found that the hypochlorite dosage needed for high-efficiency removal of DMDS resulted in carryover of chlorine odors into the scrubber exhaust stack. Obviously, this was a problem that needed to be resolved. The problem was resolved through the use of caustic and hydrogen peroxide in the third-stage scrubber for enhanced oxidation of organic compounds and depletion of chlorine residual. By using hydrogen peroxide, operators were able to increase chlorine concentration in the second-stage scrubber to make oxidation of DMDS more consistent and reliable. Enhanced solubilization of organic compounds and improved terpene removal efficiency was effected by the addition of a commercial detergent (surfactant) in the first-stage scrubber.

WSSC is currently doubling the biosolids processing capacity of the Site 2 composting facility and has started construction of another three-stage mist scrubber system (HRSD/Black & Veatch, 1993). After conducting parallel testing with the mist and packed tower systems, WSSC decided to use the mist scrubber system rather than packed towers since the mist system provided better odor control.

Operation and Maintenance Considerations (WSSC Model)

In order to maintain a high degree of performance efficiency of the modified three-stage scrubber system, WSSC incorporated an extensive daily sampling and monitoring regimen, which has led to very successful odor control. Even though the monitoring program is expensive, efforts have concentrated on identifying the specific chemical compounds responsible for odors emanating from the scrubbers. Relying on sophisticated analytical equipment such as GC/MS, Draeger tube sampling, and highly trained odor panels, specific odors have been identified and recognized.

The mist scrubber system has proven fairly easy to maintain. No problems have been reported with the open contact chambers. However, the nozzles require regular maintenance (cleaning). Nozzle maintenance in the WSSC installation is relatively easy because of redundancy; that is, multiple nozzles are installed, including spares, and can be removed and repaired or replaced without taking the entire system out of service.

In the design stage, a couple of maintenance items need to be addressed. For example, if easy access to the part or parts is not available, maintenance is difficult. Moreover, it is important to allow for equipment downtime while at the same time maintaining overall system operation. This can be accomplished if redundant pumps and blowers are included in the system.

When comparing the packed tower and the mist systems, both advantages and disadvantages become apparent. Advantages of the mist system include a lower pressure drop and elimination of packing media cleaning requirement. Disadvantages include the need for larger scrubber vessels because they rely on contact time for treatment. Moreover, due to the difficulty of removing fine droplets in the mist eliminator, an exhaust plume from the mist scrubber is more visible unless a dilution fan is used. Another disadvantage in choosing the mist system over the packed tower is higher costs, both in the initial expense for installation of water-softening and compressed-air equipment and in expenses for operation and maintenance.

ALTERNATIVE 1: MIST SCRUBBER DESIGN CRITERIA (ASP MODEL)

If decision-makers chose the mist scrubber system alternative at the ASP model composting facility, the system would probably resemble the simple line diagram shown in Figure 12.2.

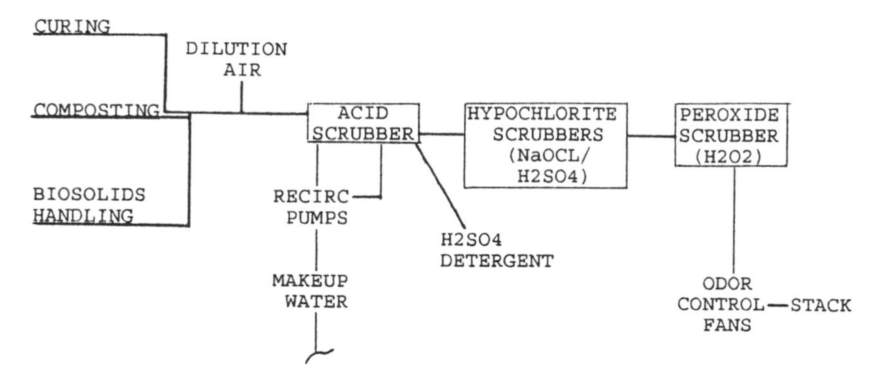

Figure 12.2 Mist scrubber system simple line diagram.

The mist scrubber system consists of four vessels as follows:

- first stage: acid wash and detergent scrubber for ammonia removal and organic compound solubilization
- second stage: two vessels ensure sufficient contact time for a second-stage sodium hypochlorite scrubber for organic sulfide oxidation
- third stage: caustic/hydrogen peroxide scrubber for enhanced oxidation of organic compounds and residual chlorine removal

To allow for equipment inspection and maintenance downtime, and to enable the system to remain in operation, bypass ducts and dampers are provided. For the chemical feed system, feed pumps, storage tanks and spill containment structures are provided for sodium hypochlorite, hydrogen peroxide, sulfuric acid, and liquid surfactants (ordinary laundry detergents). The complete system also includes a 100-foot tall discharge stack and redundant dilution fans (60,000 cfm).

ALTERNATIVE 2: PACKED TOWER SYSTEM (ASP MODEL)

In packed tower scrubber systems, used for odor control, a vertical vessel made of a synthetic material (usually fiberglass) is packed with a bed of loose plastic media. This veritcal arrangement of loose media maximizes the liquid film area for liquid/gas contact. In normal countercurrent operation, contaminated air enters below the media bed and moves upwards through the packed media bed. From a bottom sump area, chemical solution is recirculated and sprayed over the top of the media bed.

In most systems, the pH and ORP (oxidation reduction potential) of the chemical solution is continuously monitored with probes located in the pumped recirculation line or in the scrubber sump. In order to maintain pH and ORP in the optimum range for odor removal, automatic controls are usually installed. To keep impurities from building up in the recirculated liquid, a portion of the sump liquor is continuously removed.

Several manufacturers offer packed tower scrubber systems for biosolids composting facilities, including PEPCON Systems Inc., Xerxes Corporation/Heil Process Equipment and Duall Division/Met-Pro Corporation. In this text, information relating to the Duall system is used for illustrative purposes.

PACKED TOWER SCRUBBER INSTALLATION: CAPE MAY COUNTY MODEL

In an article in *Biocycle,* Goldstein (1993) points out that the Cape May County composting facility in New Jersey "has mastered the complexities of chemical scrubbing" (p. 58). For odor control, Cape May uses a three-stage packed tower system, which has evolved through several significant modifications. Augusta Fiberglass supplied the initial two-stage units, Westates the third-stage vessel.

The first-stage unit uses treated wastewater effluent with the pH adjusted to 2.5–3.5 with sulfuric acid. Treated wastewater effluent (approximately 350 gpm) is combined with 350 gpm of flow from the scrubber sump and is recirculated through the scrubber bed. This high makeup water rate is required to cool the inlet air from about 130°F to 80°F and to remove condensate from the process air. When the wastewater effluent temperature is lower (e.g., in winter), the makeup water volume can be reduced. The second-stage unit uses acidic hypochlorite (pH 6.0–6.5) with detergent (surfactant) added. The third-stage unit uses hydrogen peroxide adjusted to a pH of 8.5–9.0 with NaOH (sodium hydroxide).

The treated process air is discharged through a 100-foot tall stack without supplemental dilution air. ORP (oxidation reduction potential) is not used to control the chemical addition. Chlorine vapor is monitored by a chlorine analyzer installed in the exhaust stack. Chlorine in the vapor phase is maintained at 3–5 ppm (about 300–400 mg/L liquid concentration).

Two-stage packed tower systems are used to treat process exhaust at in-vessel composting facilities in Schenectady, New York, and Sarasota, Florida (Muirhead et al., 1993b). At one such in-vessel composting facility, a two-stage packed bed scrubber system was retrofitted. Process exhaust is treated using Duall horizontal cross-flow scrubbers for first-stage ammonia removal. Using sulfuric acid, the first stage is maintained at a pH of 2.0–2.5. The second stage used Duall vertical scrubbers with 10 feet of packing media. The chemical solution is sulfuric acid and sodium hypochlorite at a pH range of 6.5–6.9. This system incorporates a precise PLC control system with ORP and pH monitoring to control the chemical feed. Preceding the two-stage scrubber system is an air-to-air heat exchanger and a water quench tower, which removes a high percentage of the ammonia and amines along with condensate.

PACKED BED SCRUBBER PERFORMANCE

Reports on the performance of packed bed scrubber systems used in the biosolids-derived composting process have generally been favorable. For example, the Cape May County and the Schenectady systems reported excellent odor control (Goldstein, 1993; Muirhead et al., 1993b).

For composting facilities that have been in operation for several years, odor control has been an evolutionary process. As a case in point, consider that several composting facilities started operation without odor control installed or with basic two-stage systems. As time passed and neighbors began to encroach upon the composting facilities, complaints about odor began to escalate. In the late 1980s and early 1990s, many composting facilities upgraded their odor control systems. For composting facilities with two-stage systems, upgrading the system to reduce odor normally required the addition of a third stage.

Although it has been necessary for some composting facilities to upgrade to three-stage systems to improve odor control, this has not always been the case. The two-stage packed tower system at the Schenectady facility, for example, has reported good results (HRSD/Black & Veatch, 1993). The second stage was designed to treat inlet DMDS/DMS and mercaptan concentrations of 15 ppm and 10 ppm with a minimum removal of 99%. Operating data showed that with 15 ppm inlet DMDS/DMS, the system provided 99% removal. The units were designed to treat up to 400 ppm ammonia with a minimum removal of 99%. Operating data showed ammonia removal of >99.9% with 300 ppm inlet concentration.

PACKED BED SCRUBBER OPERATION AND MAINTENANCE CONSIDERATIONS

Composting and wastewater facilities that have employed packed bed scrubber systems in odor control generally report they are pleased with the performance and operation of their systems. For example, at Hampton Roads Sanitation District (HRSD), because of its extensive force main system, extensive odor-control problems were experienced with hydrogen sulfide generation and subsequent release at its nine wastewater treatment plants.

To solve this problem, HRSD used custom-fitted cover plates to cover its treatment process units and directed associated offgas into packed tower scrubbers. The primary scrubbant used in these packed towers is caustic soda. Despite minor initial problems with its scrubbers, at the present time, HRSD has several packed tower scrubbers in full operation with other units under construction. Operating with intake air hydrogen sulfide concentrations varying from 10 to 700 ppm, depending on time of year and location, and after initial problems were corrected, these scrubbers consistently achieve 98% sulfide removal (Waltrip, 1996).

The initial problems with HRSD's packed tower scrubbers included fouling of media with biogrowth and periodic decreased performance due to magnesium and calcium deposits on nozzles and packing media (McNamara, 1990). Continued pilot testing of various types of media and nozzles resulted in the installation of components that reduced these problems.

With regard to resolving media problems in packed tower scrubber systems, an approach that the Cape May County facility took required the change-out of media type. To correct problems with fouling, Cape May decided to use "self-cleaning" chevron media. It should be pointed out that the "self-cleaning" chevron media is self-cleaning to a point; that is, it must be removed once a quarter and cleaned. This process takes about four hours (HRSD/Black & Veatch, 1993).

Technological advancements in the design and operation of packed tower scrubber systems continue to be made. For example, Dupont de Nemours introduced a new type of scrubber technology in 1990, marketed under the trade name "Dynawave" by Monsanto.

Dynawave offers attractive alternatives to venturi or packed scrubbers. For example, the Dynawave scrubber is a gas/liquid contactor utilizing a "froth zone" as a mass interface. This froth zone is created by introducing the liquid stream to the gas stream countercurrently. Normally, the countercurrent action is accomplished by positioning three nozzles countercurrent to the gas stream, which, in turn, produces the froth zone. Appealing in its operational simplicity, Dynawave scrubbers can be used in place of quench towers, venturies, and packed scrubbers (McNamara, 1996).

SYSTEM DESIGN CRITERIA FOR PACKED BED SCRUBBER SYSTEM

A simple line diagram for a packed bed scrubber system is shown in Figure 12.3.

Figure 12.3 shows a three-stage system with the first stage consisting of a horizontal packed bed scrubber for ammonia removal. The horizontal scrubbers are compact units installed at ground level, which minimizes ductwork. The second and third stages are vertical towers, which use a standard 10-feet depth of packing media.

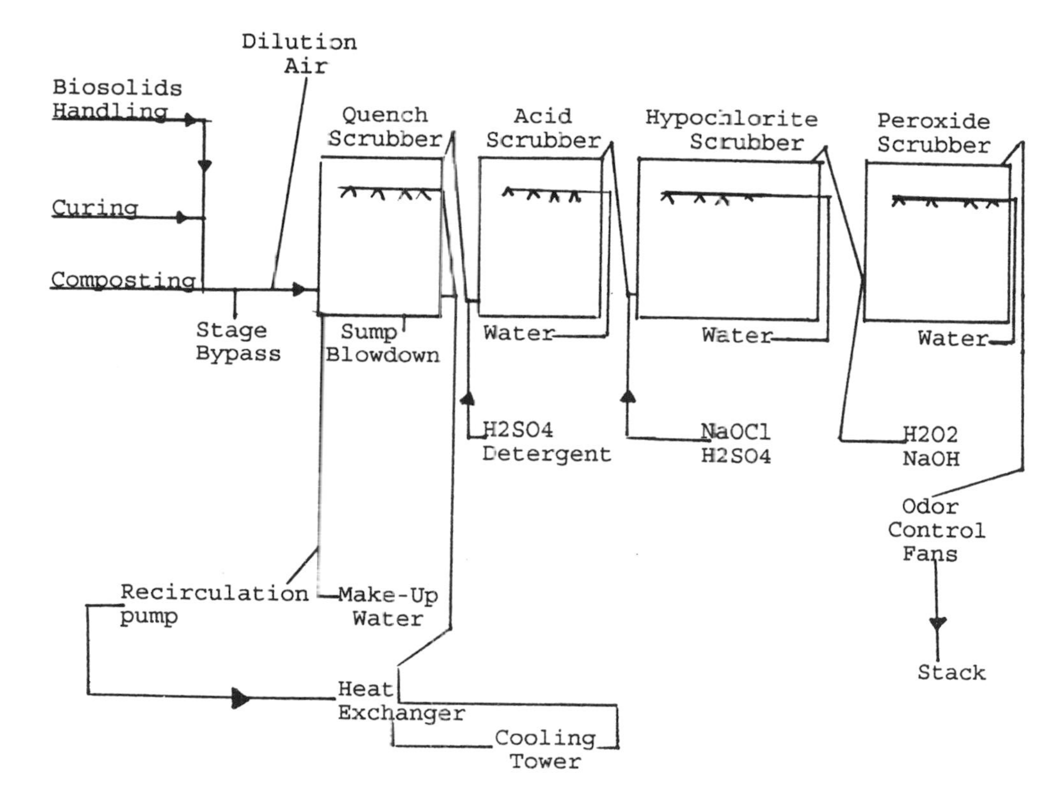

Figure 12.3 A line diagram of a packed tower scrubber system. Source: Adapted from an engineering pilot study conducted by HRSD/Black & Veatch (1993), used with permission.

Some of the system design assumptions used in this packed bed scrubber system were modified to make it more comparable to the mist system. From Figure 12.3 it can be seen that four vessels are used. Instead of a single cross-flow scrubber for the first stage, two vertical towers are used to provide cooling and ammonia removal. The second- and third-stage units use about 15 feet of packing depth in each unit to maximize contact within the vessel. An internal distribution ring is installed at the mid-depth of the packing to redistribute flow away from the vessel walls. This system includes a 100-foot dispersion stack and redundant 60,000 cfm dilution fans. This entire system was designed with flexibility in mind and could easily be adapted for the WSSC patented process.

COMPARISON OF SCRUBBER ALTERNATIVES

Table 12.1 summarizes the comparison of mist and packed bed odor-control scrubber alternatives for composting and curing air process air. When attempting to decide which one of these two alternatives is best suited for use at the composting facility, it is important to note that more extensive operating experience and performance documentation is available on the mist scrubbers used at the WSSC Site 2 and other similar biosolids composting facilities. This is not to say that the mist system is better suited for odor control at all composting facilities; rather packed bed odor-control scrubbers for biosolids composting have not been tested as extensively.

In making a decision on what type of system to install, financial considerations must also be addressed. The economics of composting, including estimated costs for both the mist and the packed bed scrubber systems will be presented later in this text.

OTHER ODOR-CONTROL ALTERNATIVES

Thus far this chapter has covered odor-control alternatives specifically dealing with scrubbers. However, other techniques have been utilized in attempting to control odor generation in the biosolids-derived composting process. Consider the example of numerous composting facilities in New England that have been using wood ash both in the mix and as an insulating cover on top of compost piles.

Reports from these plants have indicated that odors have been controlled through the use of ash. An added advantage of using ash is that the composted product does not have to be screened, screening being a source of odor emissions. The major disadvantage of using ash is that the ash cannot be reused like woodchips, which increases the production cost (Goldstein, 1993).

Whatever method is used in controlling odors generated during biosolid composting, the point is that odor generation and its control is a major consideration that cannot be overlooked. Those few biosolids composting facilities that have closed recently did so because of odor generation that ultimately resulted in complaints.

Remember, composting biosolids is a smelly process that is not a problem until the neighbors complain!

CHAPTER 13

Health and Safety Concerns

INTRODUCTION

HEALTH and safety concerns related to employees involved in the composting process and to the general public who live near the composting site are often overlooked. This is an oversight that can be very costly, not only in monetary terms related to employee medical expenses, but also in fines assessed against the responsible organization.

Other costs related to health and safety issues must also be considered. For example, insurance rates increase if health and safety concerns are overlooked or ignored. Employee lost time escalates whenever health and safety programs are lacking or nonexistent. Further public confidence in the safety of the site or in living near or doing business with the composting operation quickly deteriorates if safety and health concerns are not part of the composting facility's daily routine—part of the organization's culture. The major point made here is simple: Whatever the organization's role, whatever its mission, and whatever its vision, when an organization's culture is lacking in emphasis on safety and health, it is headed for serious problems.

Such problems can be avoided by following the guidelines specified in various safety and health regulations. As a case in point, consider the OSH Act (Occupational Safety & Health Act of 1970) and its requirements. Under this act, plant managers and other decision-makers are responsible for ensuring a place of employment free from recognized hazards that can cause or are likely to cause death or serious physical harm. In addition, management must comply with the occupational safety and health standards promulgated by the act.

The act is also specific with regard to the responsibility of the employees. Each employee must comply with the occupational safety and health standards and all rules, regulations and orders pursuant to the act that are applicable to his/her own actions and conduct.

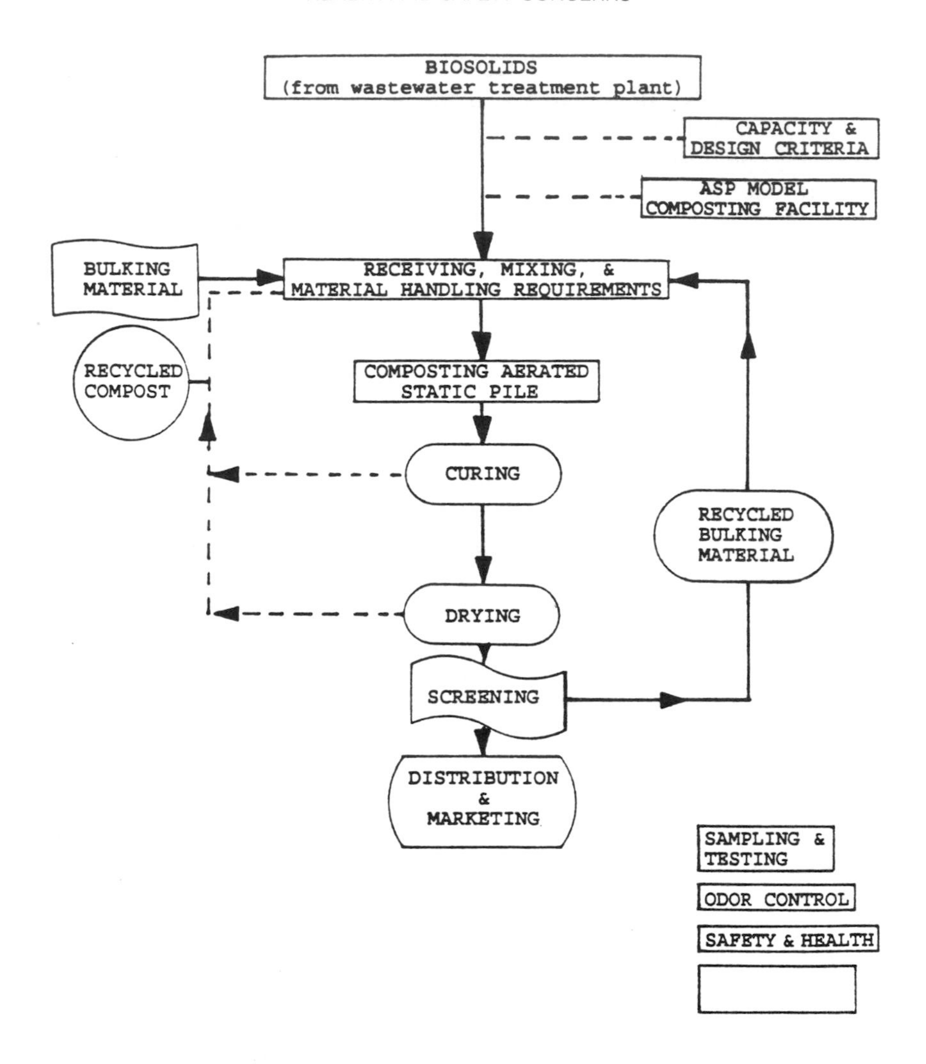

One of the major reasons why health and safety concerns have been overlooked in many industrial settings is the view by many decision-makers (managers) that "maintaining productivity" is the bottom-line; that is, productivity comes first, is foremost, and nothing else matters. This is the kind of mindset that causes workers to get injured or killed on the job.

According to another mindset, health and safety considerations are too burdensome and too expensive. No one ever said safety was inexpensive (Spellman, 1996). Incorporating health and safety programs into the organization is expensive in terms of safety equipment costs, time allocated for health and safety training, and managing health and safety programs.

However, this text takes the view that when an organization ignores health and safety considerations in the workplace, the costs are even higher! Moreover, when it comes to the burden placed on the manager in implementing health of safety programs, the manager must ask himself/herself the question: How burdensome are lawsuits for injuries received on the job, regulatory agency interference, and/or having to face the grieving family of the victim of a work-related mishap or fatality? When managers view health and safety considerations as burdensome or too expensive, another question must be asked: Burdensome and too expensive relative to what? When the decision-maker can provide an answer to this question, the organization's safety profile is doomed.

HEALTH AND SAFETY CONCERNS RELATED TO COMPOSTING

Generally, biosolids-derived composting processes are perceived to present decision-makers with problems related to nuisance conditions such as odor and dust generation. There can be little doubt that these problems are real. Nuisance conditions are created any time the composting process is exposed to the environment, improper mixing takes place, and drying operations are accomplished in excess (Haug & Davis, 1981).

Along with the inherent nuisance conditions or problems, several other composting process operations present problems. For example, the decision-maker is faced with a myriad of potential health and safety problems that he/she may not be aware of, including exposure to bioaerosols, toxic chemicals, and other substances. On-the-job injuries received while operating and maintaining composting process equipment can also be a concern.

What is the decision-maker to do about potential safety and health problems related to the composting operation? These problems can be minimized by providing adequate worker training and education. Moreover, problems related to physical hazards can be reduced in the planning, design, and construction phase. When a compost site (or any other facility) is built, it should be the goal of the designer and builder to engineer out physical hazards. For example, in the design phase, site physical hazards such as trips and falls can be avoided or engineered out of the plans and, thus, the site.

BIOAEROSOLS AND ENDOTOXINS

One of the potential health hazards related to working in a composting facility is exposure to bioaerosols. Epstein (1994) defines bioaerosols as "organisms or biological agents which can be dispensed through the air and affect human health . . . during composting, bioaerosols are not only present in waste materials but also can be generated" (p. 51).

The bioaerosols of concern during composting include bacteria, molds, actinomycetes, viruses, and fungi. However, since bacteria, molds, actinomycetes, and viruses have not been shown to be of significant importance or prevalence in biosolids-derived composting operations, it is the fungi, in particular *Aspergillus fumigatus,* that will be addressed here.

Compost facility workers and the public who live near a composting facility can be impacted by exposure to *Aspergillus fumigatus,* which has been found to grow on wood, grass, compost, rubber, and green leaves (Epstein, 1994). Specifically, *Aspergillus fumigatus* can cause *aspergillosis* in man. Aspergillosis is usually caused by inhaling air-borne spores, but can also occur through ingestion or via wounds (Singleton & Sainsbury, 1994). Man is not the only species affected by aspergillosis; it can affect a wide range of animals including horses, cattle, sheep, and poultry. The disease is an acute or chronic inflammatory infection primarily of the respiratory tract and the ear (Burnett & Schuster, 1973).

Aspergillus fumigatus is not considered a hazard to healthy individuals. Instead, it is the susceptible individuals (for example, those with chronic pulmonary function problems) who can be infected and who must, therefore, be protected. To help prevent infection, decision-makers involved with managing biosolids-derived composting facilities should incorporate a medical screening process into their hiring procedure. The procedure followed by Hampton Roads Sanitation District in pre-screening compost workers prior to their assignment at the compost site and the requirement for subsequent annual physical examinations, including annual pulmonary function testing, is an excellent model (HRSD, 1995).

With regard to protecting the public, decision-makers at the composting site should take certain measures to control generation of and exposure to the fungi. For example, when siting new facilities, the proximity to residences and public facilities should be carefully evaluated. Because there is a potential risk of increased exposure to *Aspergillus* spores in the areas located downwind from a composting site, meterological conditions that could transport bioaerosols off site should also be evaluated (Toomey, 1994).

Additional preventive measures are available. For example, a sufficient buffer area around the compost site should be incorporated. If an adequate buffer area is not feasible because of lack of available land, such measures as enclosing the process, increasing mechanization, and sound management practices can help mitigate dispersion of bioaerosols (Millner, 1995).

Table 13.1 shows locations where *Aspergillus fumigatus* are usually pres-

TABLE 13.1. Levels of *Aspergillus fumigatus* at a Biosolids Composting Facility.

Location	Concentration (Colony-Forming Units/m^2)
Mixing area	110 to 120
Near teardown pile	8 to 24
Compost pile	12 to 15
Front-end loader operation	11 to 79
Periphery of compost site	2

Source: Adapted from Epstein and Epstein (1989).

ent and the typical levels that can be expected at a biosolids composting facility.

From Table 13.1 it should be apparent that there is a direct relationship between *Aspergillus fumigatus* levels and compost site activities.

Endotoxins

Along with bioaerosols, endotoxins are also found in biosolids composting facilities (Epstein, 1994). Endotoxin is a generic term for the lipopolysaccharides of Gram-negative bacteria (Singleton & Sainsbury, 1994). Endotoxins are toxins produced within a microorganism; that is, when the cell of the microorganism is destroyed, the toxins are released. Normally, at composting facilities endotoxins are carried by airborne dust particles.

Since bioaerosols and endotoxins are both carried by dust, dust-control measures are important elements of the composting site's health and safety program. In *Composting Yard and Municipal Solid Waste* (1995), the U.S. EPA recommends incorporating dust-control measures at composting facilities to protect workers from exposure and to reduce risk of disease from these airborne hazards. The U.S. EPA (1995) lists several steps that can be taken to minimize dust generation:

- Keep compost and feedstock moist.
- Moisten compost during the final teardown, taking care not to overwet the material, which can produce leachate or runoff.
- Water roadways and surrounding pads to minimize dust.
- In enclosed operations, use engineering controls (e.g., ventilation—negative air pressure) to minimize dust.
- Isolate workers from spore-dispersing components of the composting process. Front-end loaders should have enclosed cabs that are airconditioned.
- Use aeration system's vice mechanical turning.

Employee training is essential for protecting the health and safety of compost workers from compost-generated disease-producing microorganisms. The old adage that workers cannot be expected to perform their duties in a safe manner unless they have been properly trained on the equipment/process and made aware of the associated hazards holds true in composting biosolids (Spellman, 1996). In addition, compost workers must be trained on the proper procedures for providing their own personal protection. The U.S. EPA (1995) lists the following precautions that should be taken to ensure workers' personal protection:

- Workers should wear dust masks or respirators under dry and dusty conditions. *Note:* OSHA has very specific guidelines with regard to respiratory protection for workers. One facet of OSHA's Respiratory Protection Standard that is unfamiliar to many decision-makers is the requirement that the employer must ensure that the worker is medically fit to wear a respirator. Several other requirements also apply to respiratory protection (Spellman, 1996).
- Uniforms should be provided to workers. Workers also should be informed that uniforms used in composting or other wastewater treatment plant operations should be isolated from the family laundry, otherwise, family members might be exposed. Work shoes should be handled with care as they can inadvertently expose family members to microorganisms related to composting activities.
- Education and training should be provided on the personal hygiene methods to follow when involved in composting.
- As in wastewater treatment plants, employees at composting facilities who are cut or bruised should receive prompt medical attention.
- In enclosed facilities, proper ventilation is required.

Before moving on to the next section, it should be pointed out that the purpose of this discussion is to emphasize that workers should not be assigned to work at composting facilities unless they have been cleared by competent medical authority.

TOXIC CHEMICALS (HAZARDOUS MATERIALS)

In biosolids-composting operations, the potential for exposure to volatile organic compounds (VOCs), such as chloroform and trichlorethylene, is normally not as high as when composting Municipal Solid Waste (MSW). This is not to say that biosolids-composting operations are free of toxic or hazardous materials; they are not. Because of odor-generation problems, for example, most composting operations employ some type of scrubber system, which uses hazardous chemicals/materials.

As an example, consider the wet scrubber system that uses sodium hydroxide (NaOH). Sodium hydroxide (or caustic) is a hazardous material (an alkaline corrosive) that must be used with care and caution. In concentrated solutions, sodium hydroxide can corrode glass, aluminum, lead, as well as body tissues (Meyer, 1989).

Composting operations can expose workers to other hazardous materials as well. For example, diesel fuel, gasoline, solvents, and acetylene/oxygen are chemicals often used at composting sites. Workers have a right to know what they are working with and around. Moreover, under its Hazard Communication Standard, OSHA specifically requires that employers provide workers with right-to-know information with regard to chemical hazards in the workplace (Spellman, 1996). Again, training is key in the mix of ingredients that is required to ensure worker safety on the job.

HEARING CONSERVATION

Composting operations make noise. According to OSHA, the employer has a duty to protect workers from exposure to excessive noise levels. OSHA defines noise in the workplace as "excessive" any time the level exceeds 85 DBA (DBA = A-weighted sound-pressure level). If machinery or equipment or processes generate 85 DBA or higher noise levels, OSHA requires the employer to develop a hearing conservation program. Along with being written, a hearing conservation program must include training and documentation for workers, hearing protection procedures, audiometric testing, and other requirements (Spellman, 1996).

Obviously, composting machinery that makes excessive noise should be avoided. This is not always possible, however. Equipment such as shredders, screens, front-end loaders, conveyors, blowers, and other noisy equipment are usually present in composting operations. Engineering controls can help reduce employee exposure to noise. For example, proper muffler installation on noisy equipment can help. Isolating a noisy machine, such as a blower, from the worker can also help to reduce noise levels, as can enclosing front-end loader cabs.

However, just as all noise hazards cannot be avoided, they cannot all be engineered out of a process or work site. Noise falls into a physical hazard class that is not always easy or possible to control. This shortcoming points to the need for employee training. In the real estate business the magic word is LOCATION-LOCATION-LOCATION. Similarly, in maintaining good worker health and safety the magic word is TRAINING-TRAINING-TRAINING.

Before workers are trained on the hazards of noise, it is prudent (absolutely necessary) to conduct a sound-level survey of the composting site, by a certified safety professional or industrial hygienist. During such a survey it is important that all machinery is on line to ensure that all the noise makers are indeed making noise. A map should be drawn showing the exact location of each noise maker (those that make more than 85 DBA) that is permanently installed; mobile noise makers (e.g., front-end loaders) should also be listed. Along with location, the measured noise levels should be listed for all noise makers. For example, if the composting facility uses a screening device that is not mobile, it should be drawn on the site map and its DBA noise level listed.

Hearing protection can guard workers against noise. Earmuff or ear plug-type hearing protection devices should be made available to all workers and visitors. Also, high-noise areas must be posted with appropriate warning signs.

OTHER HEALTH AND SAFETY CONCERNS

Composting operations require the use of several different types of machines. Machines are inherently dangerous, especially when they are not used in a safe manner. Even when a machine is designed to provide maximum safety to the operator, it can be dangerous whenever abused. Machine abuse can take on many forms. For example, improper maintenance can make an otherwise safe machine unsafe for use as can modifications (e.g., removal of machine guards).

When conducting maintenance on process machines, it is important to lockout/tagout the machine so it cannot be started or energized while maintenance is being performed. An OSHA Lockout/Tagout Standard specifically addresses this issue. However you can write all the safety programs you want, but if you do not provide employee training on the program and its requirements, then the written document is worthless.

This section has emphasized that worker training is an essential part of ensuring a safe composting facility. The main objectives of employee safety and health training should include the following:

- worker awareness–Workers have a right to know about the hazards they might be exposed to in the workplace.
- training–Training on correct operational procedures is critical to maintaining a safe workplace. Workers cannot be expected to perform operations safely if they have not been trained in proper operating procedures.
- PPE (Personal Protective Equipment–such as safety shoes, hardhats, safety glasses, gloves, respirators, etc.) is important. OSHA mandates that employers provide or ensure that workers

have proper PPE, know how to use PPE, and understand the PPE's limitations.
- emergency response—workers must know who to call in case of an emergency (e.g., fire); know how to safely perform basic first aid (bloodborne pathogen training is a must); and know how to contain a chemical spill (e.g., sodium hydroxide).

CHAPTER 14

The Economics of Biosolids Composting

Sound financial planning is a crucial step in the successful development of a composting program. When considering the multitude of options available for tailoring a composting program to the needs and resources of the community, decision-makers must weigh the costs and benefits involved and determine whether composting represents a feasible management option for their community. (U.S. EPA, 1995, p. 115)

INTRODUCTION

ALTHOUGH the preceding statement specifically addresses financial concerns related to the composting of yard and municipal solid waste, it is applicable to all composting operations, including biosolids-derived composting. That is, decision-makers who are involved in or are contemplating involvement in composting projects must carefully balance the economic benefits against the costs.

On the benefits side, the decision-maker should understand that sewage biosolids-derived compost is an inexpensive source of high-quality, bulk organic matter for the consumer. Commonly, the price of biosolids-derived compost is half that of peat humus and peat moss (Alexander, 1991). At the same time, however, substantial startup and production costs must be considered. For example, the decision-maker must be aware that on-site labor, bulking agent/amendment puchase, and maintainance of the aeration system, including pipe replacement, account for about 50–80% of the annual O & M (operations and maintenance) cost in the operation of a static pile composting facility like the ASP model (Goldstein, 1987).

To gain a better understanding of the cost-benefit factors and other concerns that today's decision-makers are confronted with, consider the following examples.

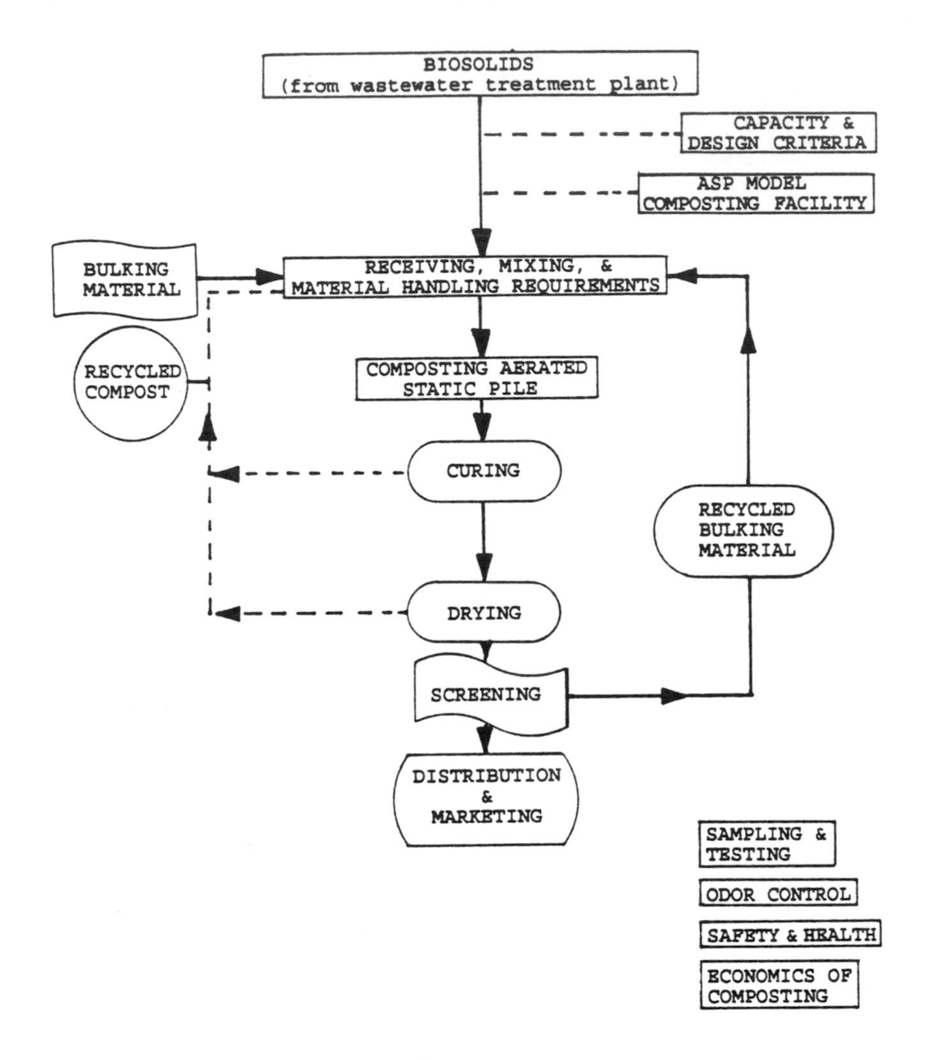

EXAMPLE ONE

Decision-makers responsible for determining cost-effective means of biosolids disposal have few alternatives. This is especially the case since ocean dumping of biosolids is no longer an ultimate disposal alternative. This prohibition only leaves a choice between the incineration and direct land-application alternatives.

If an on-line wastewater treatment facility incinerates biosolids and if the incineration process has been brought into compliance with EPA's 503 rule and other regulatory requirements, and if the presently installed system is capable of disposing of all biosolids produced, then the facility probably would choose to stay with incineration as the method of ultimate disposal. On the other hand, if this same facility has reached maximum biosolids incineration capacity, it may be looking for alternative methods of biosolids disposal.

EXAMPLE TWO

In another wastewater treatment facility, the problem may be that the presently installed biosolids incineration system is in disrepair or antiquated (cannot be upgraded to meet current regulatory standards). The system may not only be unsuited for upgrading for increased biosolids production, but might be in need of total replacement. Thus, the decision-maker needs to decide whether to replace the antiquated incinerator with a new incinerator or to replace incineration with another process such as land filling or land application.

EXAMPLE THREE

When and if the decision-maker determines that incineration of biosolids is not the best alternative, he/she must turn to the other limited alternatives. First on the list is often landfilling. If landfill space is available, and if it will remain available for future long-term use, landfilling may be the most practical alternative. Other considerations must also be taken into account. That is, if long-term landfilling is available, is it easily accessible? The decision-maker must determine how many loads and total tons of dewatered biosolids will be delivered to the landfill on a daily basis. Moreover, it must be determined how far it is from the wastewater treatment facility to the landfill. Is it cost effective to transport biosolids to a landfill site that is 85 miles from the biosolids source? In this instance, when the costs of vehicles, drivers, fuel, and maintenance costs are factored in, transporting large quantities of biosolids from the treatment plant to the landfill, a total of 170 miles roundtrip, might not be cost effective.

Along with the high transportation costs, the landfill site's tipping fees must also be considered. Are they high? When transportation costs and tipping fees are added together, are they cost prohibitive?

Yet another problem must be taken into account: the route the biosolids carrier must travel to deliver the biosolids to the landfill site. Will the biosolids trucks go through neighborhoods? Are the neighbors going to complain about truckloads of biosolids being transported through their neighborhoods? Are the present roads adequate enough to handle large trucks? Are local regulations favorable to the hauling of biosolids along the affected routes?

EXAMPLE FOUR

If it has been concluded that biosolids incineration and landfilling are not viable options, the decision-maker starts to look at the direct land-application alternative. Direct land application of biosolids has been going on throughout the United States for several years. However, disposal of biosolids on land, especially agricultural land, must be carried out in a way that provides a beneficial use without generating pollution problems associated with the accumulation of trace constituents in soils. This is the purpose, of course, of the EPA's biosolids land application regulations. The EPA's 503 rule has mandated that land-applied biosolids must meet several requirements. *Note:* Generally, this has not been a problem for biosolids producers, but it must be considered.

Assuming that the decision-maker's treatment plant(s) produces biosolids of "Exceptional Quality" and otherwise suitable for land application, several questions must be answered prior to starting such a program. One of the first questions is whether or not agricultural, forest, and/or disturbed land areas are available for direct land application of biosolids. If application sites are available, how far from the biosolids production site are they?

Probably the most important question that must be answered has to do with local mindset. That is, are the potential customers (farmers, mining company representatives, and forestry services managers) receptive to the idea of using biosolids in land application? The decision-maker may have a better chance of winning the argument for direct land application of biosolids by pointing out that long-term recycling of biosolids will benefit all the parties involved. This is especially the case when biosolids is applied properly according to present guidelines and regulations. This is a point, of course, that cannot be overemphasized.

After gaining positive answers to these questions, the decision-maker must also determine if there is a site large enough or enough smaller sites locally to receive all the biosolids product generated by the treatment plants in the area.

If positive answers to all of the preceding questions can be obtained, the decision-maker may chose land application of biosolids as the disposal alternative.

EXAMPLE FIVE

In this final example, the decision-maker may have found, through research, that none of the alternatives mentioned previously is acceptable. On the other hand, maybe some or all of the preceding alternatives are acceptable and practical, but only in limited fashion. That is, the decision-maker may determine that approximately 25% of the generator's biosolids can be land applied. Along with land application, maybe 60% of the biosolids can be incinerated. But what about the remaining 15%?

This is where biosolids-derived composting normally enters into the picture. The decision-maker has 15% of the biosolids left over and must decide what to do with it. Since all the disposal alternatives looked at to this point have been utilized to the maximum or rejected, composting is the only possibility left.

Deciding to process biosolids for ultimate disposal/recycling is the easy part of the problem. Tougher questions remain that must be asked and answers found for. For example, is a suitable site available to build a compost facility? If such a site is available, will the neighbors accept it? How much will it cost to build a composting facility? How much will it cost to properly equip a composting facility? How much will it cost to operate and maintain a composting facility? Is the compost produced to be given away, or is it going to be sold?

This last question is critical. If the compost is to be given away, to whom? Are there end users who want compost? If the compost product is to be sold, other questions present themselves. Consider the argument made by Rhonda Oberst (1996), recycling manager/agronomist for Hampton Roads Sanitation District, that the key to success for a composting facility is marketing: Without a market, what is to become of the compost? Many decision-makers forget to ask themselves this vital question.

With regard to marketing compost, some decision-makers might answer the question about whether or not a market exists for biosolids-derived compost without further investigation. This is a mistake. Marketing questions should be answered by those who are expert in marketing. As a case in point, consider the following case study.

CASE STUDY

A decision-maker for a well-known utility who is responsible for wastewater treatment and biosolids disposal is confronted with the problem of disposal of biosolids. After looking at all the disposal options, she finds that most of them are not feasible. Then, out of the clear blue sky the decision-maker hits on the solution: biosolids-derived composting.

This decision might be the right decision. But there is a problem. The decision-maker has merely used conjecture; she imagines that the composting of biosolids is the "golden goose"—it will make her a legend among her peers—an innovator without equal—a money-making machine. Composting biosolids and then selling the product to the local public will make exorbitant amounts of money for her utility. The decision-maker is so excited about the prospects of turning a waste product into that magic "green" stuff (not to be confused with agricultural green, such as soybeans, wheat, corn, tomatoes, etc.) that she leaps into the construction of the biosolids composting facility for her utility, without any further considerations.

You don't need to be an economist to figure out the moral of this case study.

This chapter presents information that will help prevent decision-makers from making decisions about starting a biosolids-derived composting process without proper planning. The key word is "planning." Without planning, any biosolids-derived composting operation is almost certain to fail. The decision-maker's preplanning efforts must preclude failure. To help the decision-maker avoid failure, the following sections are presented.

THE DECISION-MAKING PROCESS

The information presented in the following sections is based on several assumptions in an effort to assist the decision-maker in determining whether or not biosolids composting is a viable cost-benefit option.

Costs are site specific (Haug, 1980). Moreover, as stated earlier, the major factors affecting operating costs are labor, bulking material cost, and O & M costs. In addition, recovery product value and remuneration must be a part of the cost-benefit equation (Corbitt, 1990). While it is true that the amount of compost currently produced is relatively small in comparison to the potential demand, it is also true that "decision-makers should not expect to earn money from composting" (EPA, 1995, p. 115). The point is that the decision-maker must take into account that a biosolids composting operation is going to accrue expense that must be borne by other accounts. With these key points, in mind, the following assumptions are presented.

ASSUMPTIONS

(1) The biosolids-derived composting process described here, in relation to costs, is the aerated static pile method (ASP model) based on data provided by Finley (1996) and Hampton Roads Sanitation District/ Black & Veatch (1993). It is based on typical numbers and not on any particular site.
(2) The dewatered biosolids delivered to the ASP model composting facility has a solids concentration of 20–25%.
(3) The bulking agent (or amendment) used is hardwood chips.
(4) The ASP model composting facility is rated at 12 to 17.5 dry ton/day capacity.

(5) Approximately 17,500–25,000 cubic yards of finished compost product is marketed on an annual basis (all finished compost product produced is sold to end users).

(6) The cost of composting ranges from $275 to $300/dry ton and is competitive with incineration, direct-land application, and other methods employed to manage biosolids (these range from $180 to $400/dry ton.

(7) The finished compost product meets the U.S. Environmental Protection Agency's part 503 "Exceptional Quality" limits for metals and Class A requirements for pathogens and vector attraction reduction.

(8) The finished compost product is sold at $15/cubic yard for small loads (standard 1/2 ton pickup truck-bed size) and at $10/cubic yard for dumptruck-size and larger.

(9) Total sales per annum offset operational costs by 20%.

(10) Approximately 25,000 bags of compost are sold annually, accounting for 5% of sales by volume. Bag sales account for a larger portion of overall revenue, because packaging increases the value to $28/cubic yard.

(11) For aeration of active compost piles, three independent, adjustable computer-controlled blowers maintain aeration.

(12) The ASP model is installed on a 50-acre site with 250 feet of surrounding buffer region.

(13) All construction, equipment, odor-control, personnel, and operating costs presented in this model are estimates of typical costs projected by vendors, contractors, engineering consultants and on previous project experience. Costs for site development such as water and sewer extensions, access roads, clearing, and rough grading are not included in the summary cost-estimate outline presented in Table 14.1. In addition, the cost-estimate outline does not include costs for land purchase, engineering, legal, administration, and other associated items. All costs reflect installed prices for the individual items in question.

GENERAL REQUIREMENTS COSTS

General requirements costs include the general contractor's overhead and profit. In addition, cost burdened to the contractor, such as insurance and bonding, equipment rental, scheduling, mobilization, and temporary facilities and utilities is included. The costs for general requirements account for 8% of the project construction cost.

TABLE 14.1. Summary Cost Estimate Outline.

Facility Component	Cost ($-Rounded)
General requirements (8%)	1,106,000
Sitework	1,765,000
Compost process building and equipment	6,220,000
Administration building	245,000
Maintenance building	240,000
Compost storage building	1,037,000
Bagging building and equipment	85,000
Odor-control system	1,985,000
Miscellaneous	160,000
Material handling equipment	1,609,000
Electrical, power and lighting	992,000
Subtotal cost	15,444,000
Contingency allowance	1,378,000
Total construction cost	16,822,000

SITEWORK

The costs of concrete pavement surrounding structures on site, asphalt parking, finished grading and seeding, on-site water and sewer lines, security fencing, drainage ponds associated with pavement and roof-top runoff, and erosion and sediment control during the construction process are all included in the sitework account. Concrete pavement costs were estimated based on a 10-inch slab thickness. Moreover, all concrete costs include the cost of excavation, sub-base, compacting, forming, reinforcement, placing and finishing of the concrete. Miscellaneous sitework costs such as stormwater and wastewater facilities and fencing were estimated from design (conceptual) layouts.

COMPOST PROCESS BUILDING AND EQUIPMENT

The compost process building is preengineered with all structural steel galvanized to minimize corrosion and maintenance. A 10-inch thick concrete slab and footings are included in the costs. Retaining walls, which vary in height depending on their use, are 18 inches thick. This cost account also includes a precast concrete aeration trench, which, along with the process building slab, will be sloped. Screening equipment and ancillary facilities costs are included in this account.

OTHER BUILDINGS

The administrative, maintenance, and bagging buildings costs are based on square footage as related in the following: administrative (45′ × 75′), maintenance (60′ × 100′), and bagging (45′ × 65′). Concrete slab and footing costs were included in the concrete costs. Costs for heating and ventilation for all buildings except the compost process and storage buildings are also included. Only the maintenance and administrative buildings have costs factored in for plumbing and fixtures. Additional costs for equipping the maintenance building with two overhead doors for front-end loader maintenance were factored in. Costs for equipment required to outfit the administration, maintenance, and bagging buildings such as the cost of desks, chairs, cabinets, tools, microcomputers, repair parts, and bagging equipment are also included in this account.

ODOR-CONTROL SYSTEM

The cost estimate for the odor-control system is based on installation of a three-stage mist scrubber system typically used for such installations. Chemical usage and cost information was based on information obtained from two mist scrubber operating units, Lancaster, PA, and WSSC Site 2. The actual chemical costs for the odor-control system at the ASP model composting facility are approximately $175,000. Other annual operating costs for the odor-control system include electrical power at approximately $110,000, operator labor costs at $27,500, and normal preventive maintenance costs of $16,000.

MISCELLANEOUS

Miscellaneous costs include the cost of a weigh scale, a fuel station for diesel and gasoline (based on a 5,000-gallon diesel fuel tank and 1,000-gallon gasoline tank), a 20 × 30′ canopy, and a fuel management system.

MATERIAL HANDLING EQUIPMENT

This account includes the cost of six front-end loaders, one forklift, and one streetsweeper.

ELECTRICAL COSTS

Electrical costs are based on 10% of equipment and facility construction costs.

CONTINGENCY ALLOWANCE

A contingency allowance of approximately 10% has been added to the subtotal cost for items that may change during the design process.

OPERATION AND MAINTENANCE (O & M) COSTS

After the compost facility construction has been completed, after composting equipment has been procured and installed, and after all required permits have been obtained (if necessary), the composting facility is ready to be staffed and placed in operation. However, it is important for the decision-maker to have prior knowledge of the approximate costs related to operations and maintenance of composting.

To facilitate this critical preplanning phase, this section will present information about operating and maintenance expenses that can be expected for a composting facility based on the ASP model. The figures presented in Table 14.2 are typical costs; they do not reflect the budgeting figures of any on-line operation in the United States. Instead, they reflect a balance between high- and low-cost operations related to a nonunionized and a unionized workforce, costs that are lower and higher depending upon location in the United States, and other costs that may be higher or lower depending upon suppliers [e.g., costs for bulking agents vary considerably depending upon where they are procured and other competing entities (pulp mills) that may drive up costs].

PERSONNEL SERVICES

The account for personnel services includes costs for salaries and wages, overtime, licensing bonuses, part-time help, and temporary service help. The costs for salaries and wages, overtime, and licensing bonuses are based on a fulltime workforce of ten personnel. The ASP model composting facility is staffed as follows: (1) plant superintendent, (1) administrative assistant/sales clerk, (3) compost facility operators, (1) diesel mechanic,

TABLE 14.2. Projected Operating and Maintenance Expenses.

Object Description	Approximate Annual Cost
Personnel services	$335,000
Fringe benefits	$83,000
Materials and supplies	$44,000
Transportation	$40,000
Utilities	$7,000
Bulking agent	$193,000
Contractual services	$12,000
Miscellaneous services	$27,000
Total	$741,000

(1) maintenance operator, (1) compost facility operator assistant, and (2) driver operators for a total of ten staff members.

Licensing bonuses are added to workers' hourly wages. For example, if a compost worker goes from a Wastewater Class IV license to a Class III and above, he/she receives a slight increase in hourly wage. Not all composting facilities follow this licensing bonus procedure, but for those who do (e.g., HRSD), these small bonuses have proven beneficial to the worker and the organization.

As a case in point, consider the worker at the composting facility who is interested in advancing in the wastewater industry to management level positions. This is difficult if it were not for the provision allowing compost workers to work in wastewater treatment plants long enough to gain the required experience; that is, to establish their eligibility for a wastewater operator's license.

FRINGE BENEFITS

The fringe benefit account is based on ten fulltime employees and specifically lists the costs for FICA, retirement, group life insurance, hospitalization, and uniform and safety shoe allowance.

MATERIALS AND SUPPLIES (M & S)

The M & S account summarizes the total projected costs for general supplies, electrical components and labor, lubricants, housewares, and safety equipment.

TRANSPORTATION

The transportation account includes funding for personnel travel expenses and mileage, gasoline, diesel fuel, and equipment maintenance.

BULKING AGENT

The costs associated with procuring bulking agents for composting are site specific, or more correctly stated, regionally specific. According to Oberst (1996), hardwood chips come at a premium—about $20 to $27/ton—partly because Oberst's facility is competing with paper mills in the region. However, these chips can be reused about three times in the process, which saves money in the long run.

CONTRACTUAL SERVICES

This account includes costs for general contractual services such as photocopy and repairs, for electrical repair to aeration blowers and control equipment, and research projects.

MISCELLANEOUS EXPENSES

The miscellaneous account includes the cost of marketing, packing, and advertising the compost product. Expenses involved in professional training of employees and fees for professional organizational membership are also included here.

SUMMARY

Based on information compiled by Goldstein and Steuteville (1993), an estimated 1 milion dry tons (900,000 dry metric tons) of biosolids were composted and recycled in the United States in 1993. Slivka (1992) projects that the potential demand for biosolids-derived compost will increase from about 35 million tons per year to more than 450 million per year in the near future. Borberg and McLemore (1983) point out that compost marketed by HRSD to private individuals, nurseries, and landscapers has shown steady growth. Moreover, Finley (1996) states that he cannot produce enough biosolids-derived compost to meet the demand.

From the information provided above, it seems clear that biosolids-derived composting has a rosy future. Moreover, with increased awareness (gained through active marketing and educational programs) of the benefits of using biosolids-derived compost in agriculture, silviculture, and sod production, the biosolids-derived composting process should continue to grow as a beneficial end-use disposal method.

This chapter (and the entire text for that matter) has attempted to lay out in a very basic and understandable form the need to assess the long-term feasibility of any potential composting operation. Such a feasibility study should focus on identifying all costs associated with the operation. The desision-maker must keep in mind that composting does not make money; that is, the cost of producing compost exceeds the market value of the product. However, the decision-maker should look at composting as a viable alternative to incineration and landfilling and make a careful comparison of the costs of landfilling, incinerating, and composting. In many instances, composting can compete favorably with landfilling and incineration as the ultimate disposal method.

Probably the best way to gain a better understanding of the cost-related preplanning that should be done by decision-makers before launching into a full-blown composting project is to observe the advice given by the U.S. EPA (1995). Although the following advice focuses on composting yard and municipal solid waste, it is pertinent to all types of composting operations.

> The cost components of the various composting systems are the major determinants in choosing a composting system. Judging whether a composting program will save money is difficult and depends as much on local circumstances as on the chosen combination of collections and processing. A municipality's size in proportion to its labor costs, and equipment cost and operating rates will determine much of its composting costs. While it is impossible to consider every contingency, planners must approach the issue of costs and benefits from this perspective, drawing all relevant factors into the equation to make a sound decision on composting in their community. To determine the savings and thus the economic feasibility of a composting facility, planners should evaluate the cost per ton of material composted and compare these numbers with the costs of alternative management options. (p. 120)

References

Albrecht, R. (1987). How to succeed in compost marketing. *BioCycle,* September, 28 (9) 26–27.

Alexander, R (1991). Sludge compost use on athletic fields. *BioCycle,* July, 32 (7) 69–71.

American Public Health Association (APHA) (1992). *Standard Methods for the Examination of Water and Wastewater* (18th ed.) Washington, DC.

Baumler, R. (1996). Personal Communication. January 5, 1996.

Benedict, A. H., Epstein, E., & English, J. N. (1986). Municipal sludge composting technology evaluation. *Journal WPCF,* April, 58 (4) 279–289.

Borberg, J. R., & McLemore, M. (1983). Regional compost facility serves Virginia communities. *BioCycle,* May/June, 24 (3) 19–21.

Burnett, C. H. (1992). Small cities + warm climates = windrow composting. Presented at the *Water Environment Federation 65th Annual Conference & Exposition,* New Orleans, LA. September, 20–24.

Burnett, G. W. & Schuster, G. S. (1973). *Pathogenic Microbiology.* St. Louis: C. V. Mosby Company.

Cheremisinoff, P. N. (1995). Gravity separation for efficient solids removal. *The National Environmental Journal,* Nov/Dec, 5 (6) 29–32.

Cheremisinoff, P. N., & Young, R. A. (1981). *Pollution Engineering Practice Handbook.* Ann Arbor, MI: Ann Arbor Science Publishers, Inc.

Corbitt, R. A. (1990). *Standard Handbook of Environmental Engineering.* New York: McGraw-Hill, Inc.

Davis, M. L., & Cornwell, D. A. (1991). *Introduction to Environmental Engineering* (2nd ed.). New York: McGraw-Hill, Inc.

DeBertoldi, M., Citernesi, U., & Griselli, M. (1980). Bulking agents in sludge composting. *Compost Science & Land Utilization,* January/February, 21 (1) 32–35.

Diaz, L. F., Savage, G. M., Eggerth, L. J., & Golueke, C. G. (1993). *Composting and Recycling Municipal Solid Waste.* Boca Raton, FL: Lewis Publishers.

Epstein, E. (1994). Composting and bioaerosols. *BioCycle,* January, 35 (1) 51–58.

Epstein, E., & Alpert, J. E. (1984). Sludge dewatering and compost economics. *BioCycle,* 25 (10) 31–34.

Epstein, E., & Epstein, J. (1989). Public health issues and composting. *BioCycle,* August, 30 (8) 50–53.

Finley, D. (1996). Personal Communication. January 5, 1996.

Finley, D., & Morse, D. (1996). Personal Communication. January 5, 1996.

Finstein, M. S., Miller, F. C., Hogan, J. A., & Strom, P. F. (1987). Analysis of EPA guidance on composting sludge. *BioCycle,* January, 28 (1) 20–26.

Finstein, M. S., Miller, F. C., & Strom, P. F. (1986). Monitoring and evaluating composting process performance. *Journal of WPCF,* 58 272–278.

Fitzhugh, M., Norton-Arnold, & Fischer, V. (1994). Examining markets for biosolids. *BioCycle,* June, 35 (6) 72–73.

Goldstein, N. (1985a). Learning from experience. *BioCycle,* January/February, 26 (1) 22–25.

Goldstein, N. (1985b). Sewage sludge composting facilities on the rise. *BioCycle,* November/December, 26 (8) 19–24.

Goldstein, N. (1987). Technology evaluation at compost sites. *BioCycle,* May/June, 28 (3) 28–33.

Goldstein, N. (1993). EPA releases final sludge management rule. *BioCycle,* January, 34 (1) 56–59.

Goldstein, N., Riggle, D., & Steuteville, R. (1994). Biosolids composting strengthens its base. *BioCycle,* December, 35 (12) 48–57.

Goldstein, N., & Steuteville, R. (1993). Biosolids composting makes healthy progress. *BioCycle,* December, 34 (12) 48–57.

Golueke, C. G., & Diaz, L. F. (1987). Composting and the limiting factor principle. *BioCycle,* April, 28 (4) 22–25.

Gray, K. R., Sherman, K., & Biddlestone, A. J. (1971). A review of composting: Part I. *Process Biochemistry,* 6 (6) 32–36.

Haller, E. J. (1995). *Simplified Wastewater Treatment Plant Operations.* Lancaster, PA: Technomic Publishing Co., Inc.

Hampton Roads Sanitation District (1981). Aspergillus fumigatus: *A Background Report.* Virginia Beach, VA.

Hampton Roads Sanitation District (1990). *Odor Control Engineering Pilot Study: Peninsula Composting Facility.* Newport News, VA.

Hampton Roads Sanitation District (1993). *Pilot Engineering Study Involving Upgrading Present Composting Facility with New State of the Art Odor Control Devices.* Virginia Beach, VA.

Hampton Roads Sanitation District (1995). Safe work practices for compost site operators. *HRSD's Safe Work Practices.* Virginia Beach, VA.

Hampton Roads Sanitation District/Black & Veatch (1993). *Preliminary Engineering Report for Peninsula Composting Facility.* Virginia Beach, VA.

Haug, R. T. (1980). *Compost Engineering: Principles and Practices.* Lancaster, PA: Technomic Publishing Co., Inc.

Haug, R. T. (1986). Composting process design criteria: Part III. *BioCycle,* October, 27 (10) 53–57.

Haug, R. T., & Davis, B. (1981). Composting results in Los Angeles. *BioCycle,* November/December, 22 (6) 19–24.

Henry, C. L., & Harrison, R. B. (1992). Comparing yard waste and sludge compost. *BioCycle,* February, 33 (2) 42–47.

Higgins, A. J. (1983). Reducing cost for bulking agents. *BioCycle,* September/October, 24 (5) 34–38.

Higgins, A. J., Kasper, V., Derr, D. A., Singley, M. E., & Singh, A. (1981). Mixing systems for sludge composting. *BioCycle,* September/October, 22 (5) 18–22.

Higgins, A. J., Suhr, J. L., Rahman, M. S., Singley, M., & Rajput, U. S. (1986). Shredded rubber tires as bulking agent in sewage sludge compost. *Waste Management and Research,* 4 (4) 367–386.

Laws, E. A. (1993). *Aquatic Pollution* (2nd ed.). New York: John Wiley & Sons, Inc.

Lester, F. N. (1992). Sewage and sewage sludge treatment. In R. Harrison (Ed.) *Pollution: Causes, Effects, & Control.* London: Royal Society of Chemistry (pp. 33–62).

Lewicki, C. (1996). Personal Communication. February 16, 1996.

Lue-Hing, C., Zenz, D. R., & Kuchenrither, R. (1992). *Municipal Sewage Sludge Management: Processing, Utilization, and Disposal.* Lancaster, PA: Technomic Publishing Co., Inc.

Masters, G. M. (1991). *Introduction to Environmental Engineering & Science.* Englewood Cliffs, NJ: Prentice Hall.

McGhee, T. J. (1991). *Water Supply and Sewerage.* New York: McGraw-Hill, Inc.

McNamara, B. (1996a). *Compost Odor Control Study.* Virginia Beach, VA: Hampton Roads Sanitation District.

McNamara, B. (1996b). Personal Communication. January 5, 1996.

M'Coy, W. S., Haley, M. A., & Jain, A. C. (1994). Centrifuge considerations. *Water Environment & Technology,* October, 6 (6) 52–56.

Mendenhall, T. C. (1990). Public acceptance strategies for sludge utilization. *BioCycle,* October, 31 (10) 34–37.

Metcalf & Eddy. (1991). *Wastewater Engineering: Treatment, Disposal, & Reuse* (3rd ed.). New York: McGraw-Hill, Inc.

Meyer, E. (1989). *Chemistry of Hazardous Materials* (2nd ed.). Englewood Cliffs, NJ: Prentice Hall, Inc.

Millner, P. (ed.) (1995). Bioaerosols and composting. *BioCycle,* January, 36 (1) 48–54.

Moran, J. M., Morgan, M. D., & Wiersma, J. H. (1986). *Introduction to Environmental Science* (2nd ed.). New York: W. H. Freeman & Company.

Muirhead, T., LaFond, P., & Buckley, S. (1993a). Odor control for biosolids composting. *BioCycle,* 34 (2) 64–70.

Muirhead, T., LaFond, P., & Dennis, D. (1993b). Air handling and scrubber retrofits optimize odor control. *Biocycle,* 34 (3) 68–75.

Oberst, R. (1996). Personal Communication. January 5, 1996.

Oberst, R., & Robinson, B. P. (1995). Beneficial reuse of biosolids. Paper presented at *WEFTEC '95,* Miami.

Odette, R. (1995). Winning public approval. *Water Environment & Technology,* May, 1995.

Outwater, A. B. (1994). *Reuse of Sludge and Minor Wastewater Residuals.* Boca Raton, FL: Lewis Publishers.

Padmanabha, A., Locke, E. R., Bailey, W. F., & Tolbert, D. A. (1994). Solids processing upgrade challenges Blue Plains. *Water Environment & Technology,* July, 6 (7) 51–56.

Peavy, H. S., Rowe, D. R., & Tchobanoglous, G. (1991). *Environmental Engineering.* New York: McGraw-Hill, Inc.

Rynk, R., et al. (1992). *On-Farm Composting Handbook.* Ithaca, NY: Cooperative Extension, Northeast Regional Agricultural Engineering Service.

Shea, T. G., Braswell, J., & Coker, C. S. (1980). Bulking agent selection in sludge compost facility design. *Compost Science/Land Utilization,* November/December, 21 (6) 20–21.

Shimp, G., Bond, M., Sandino, J., & Oerke, D. (1995). Budgets improving bottom lines with solids thickening and dewatering. *Water Environment & Technology,* 7 (11) 44–49.

Singleton, P., & Sainsbury, D. (1994). *Dictionary of Microbiology & Molecular Biology* (2nd ed.). New York: John Wiley & Sons.

Slivka, D. C. (1992). Compost: United States supply & demand potential. *Biomass & Bioenergy.* Tarrytown, NY: Pergamon Press. 3 (3–4) 281–299.

Soil Science Society of America (SSSA) (1994). *Sewage Sludge: Land Utilization & the Environment.* Madison, WI: SSSA.

Sopper, E. A. (1993). *Municipal Sludge Use in Land Reclamation.* Boca Raton, FL: Lewis Publishers.

Sopper, W. E., & Kerr, S. N. (1981). *Revegetating Strip-Mined Land with Municipal Sewage Sludge.* Project Summary USEPA Report 600/52-81-182. Washington, DC: Government Printing Office.

Spellman, F. R. (1996). *Safe Work Practices for Wastewater Treatment Plants.* Lancaster, PA: Technomic Publishing Co., Inc.

Spencer, R. (1992). Sludge composting takes town out of landfill. *BioCycle,* January, 33 (1) 52–54.

Spohn, E. (1970). Composting by artificial aeration. *Compost Science,* May/June.

Sundstrum, D. W., & Klei, H. E. (1979). *Wastewater Treatment.* Englewood Cliffs, NJ: Prentice-Hall, Inc.

Tarr, J. A. (1981). City sewage and the American farmer. *BioCycle,* September/October, 22 (6) 36–38.

Tchobanoglous, G., Theisen, H., & Vigil, S. A. (1993). *Integrated Solid Waste Management.* New York: McGraw-Hill, Inc.

Tivy, J. (1990). *Agricultural Ecology.* New York: McGraw-Hill, Inc.

Toomey, W. A. (1994). Meeting the challenge of yard trimmings diversion. *BioCycle,* May, 35 (5) 55–58.

Turk, J., & Turk, A. (1988). *Environmental Science* (4th ed.). Philadelphia: Saunders College Publishing.

U.S. Army Corps of Engineers (1992). *The Chesapeake Bay Program Erosion Control Study,* p. 181.

U.S. Environmental Protection Agency (EPA) (1982). *Dewatering Municipal Wastewater Sludges,* EPA-625/1-82-014. Cincinnati: Center for Environment Research Information.

U.S. Environmental Protection Agency (EPA) (1993). *Standards for Use or Disposal of Sewage Sludge,* Final rule, 40 CFR Part 503. *Federal Register 58* (32):9248–9415. 19, February 1993, Washington, DC: U.S. Government Printing Office.

U.S. Environmental Protection Agency (EPA) (1994). *Biosolids Recycling: Beneficial Technology for a Better Environment,* EPA 832-R-94-009.

U.S. Environmental Protection Agency (EPA) (1995). *Composting Yard and Municipal Solid Waste.* Lancaster, PA: Technomic Publishing Co., Inc.

U.S. Department of Labor (1995). 29 CFR part 1910, *Occupational Safety & Health Standard for General Industry.* Chicago: CCH Incorporated.

Vesilind, P. A. (1980). *Treatment and Disposal of Wastewater Sludges* (2nd ed.). Ann Arbor, MI: Ann Arbor Science Publishers, Inc.

Waltrip, G. D. (1996). Personal Communication. January 5, 1996.

Water Environment Federation (WEF) (1995). *Biosolids Composting.* Alexandria, VA: WEF.

Water Environment Federation (WEF), & American Society of Civil Engineers (ASCE) (1995). *Odor Control in Wastewater Treatment Plants.* Alexandria, VA: WEF & ASCE.

Wilbur, C., & Murray, C. (1990). Odor source evaluation. *BioCycle,* 31 (3) 68–72.

Williams, T. O. (1995). Odors & VOC emissions control measures. *BioCycle,* May, 36 (5) 49–56.

Index